Hydatid Disease

With Special Reference to Kenya

Hydatid Disease

With Special Reference to Kenya

G.B.A. Okelo

A.K. Chemtai

C. Marcus French

Nairobi University Press

First Published 1991 by
Nairobi University Press
University of Nairobi
P.O. Box 30197
Nairobi

University of Nairobi Library
Cataloguing in Publication Data

Okelo, G.B.A.
    Hydatid Disease : with special reference to Kenya / G.B.A. Okelo,
A.K. Chemtai and C.M. French. Nairobi: Nairobi University Press, 1991.

1. Echinococcus.   I. Chemtai, A.K.   II. French, C.M.   III. Title

RC186.T6038

ISBN 9966 846 18 2

Printed by College of Education and External Studies
University of Nairobi
P.O. Box 30197
Nairobi

Table of Contents

List of Figures

List of Tables

Foreword

This book discusses an interesting medical and veterinary problem in Kenya. The book concentrates on medical aspects of hydatid disease, including the transmission of the disease by dogs from domesticated animals such as cattle and camels to man.

Hydatid disease is highly prevalent among the Turkana due to various cultural habits which will be discussed in this book. The disease occurs as painless abdominal swellings, which can be extremely large before the patients seek medical treatment. This problem is compounded by the fact that Turkana District has not been accesssible and was neglected in the colonial days. With improved transportation, these long delays before individuals seek medical attention will decrease.

The dog, "man's greatest friend", is responsible for the transmission of hydatid disease among the Turkana. Because of the close contact of dog and man, the dog which is left to eat infected beef, camel and even human corpses, transmits the disease to man right from early childhood. The dog is the target in the control of the disease.

The book has six chapters, each set out to discuss in full an aspect of hydatid disease. In order to understand how the disease is prevented among the Turkana people, the authors have ably studied the lifestyles of these people. This knowledge is a prerequisite to implementation of preventive strategies. Prevention and treatment are described in Chapter 5 and draw heavily on a cultural approach.

The authors have drawn on the experiences of others in the historical review which is necessary as a prelude to their own research and clinical work. The authors then describe their experiences in the chapter on clinical aspects and diagnosis. Definite diagnosis using immunodiagnostic methods is widely covered.

The Government has embarked on a programme to eliminate stray dogs in Turkana District in the effort to control hydatid disease. This effort should be supplemented by health education aimed at changing some of the people's habits.

N.O. Bwibo

Professor of Paediatrics

September, 1991

Preface

This book is based on our experiences on hydatid disease in Kenya. We focus on the situation in Turkana: its people, environment and the worm *Echinococcus granulosus*. Examples are also drawn from other areas of Kenya for comparison.

The book covers important areas of human hydatidosis: epidemiology, immunology, diagnosis and treatment including control programmes. The customs and traditions of the Turkana are described in depth, especially those aspects related to the transmission and treatment of hydatid disease.

The information provided gives a good understanding of hydatid disease as a worldwide public health problem. A wide section of audience, particularly medical students, doctors, scientists, social workers, public health inspectors and clinical officers will greatly benefit from various aspects of this book.

Suggestions, criticisms and contributions are welcome for incorporation in future editions of the book.

G.B.A. Okelo

C.M. French

AK. Chemtai

November, 1991

Preface

[illegible] [illegible] [illegible] [illegible] [illegible] [illegible] 1986 [illegible] Workshop on [illegible] [illegible] in Degree [illegible] courses [illegible] work [illegible] book a useful [illegible].

[illegible] [illegible] [illegible] [illegible] further reading [illegible] guidance to [illegible] [illegible] assignments and [illegible] examples [illegible] programmed. The [illegible] [illegible] treatment [illegible].

[illegible] foundation [illegible] [illegible] extended treatment of several case studies [illegible].

[illegible] [illegible] and contributions of [illegible] [illegible].

C. [illegible]
C. [illegible]
J.A. [illegible]

November 1991

Introduction

Hydatid disease is a major public health problem with the highest endemicity in man in the world. The two endemic foci in Kenya are Turkana district and Masailand. The economic importance of the disease has not yet been fully estimated. However, a lot of meat which is a source of animal protein for man in Kenya is being thrown away annually at abattoirs. This is particularly serious in dry parts of Kenya which have little food available; thus exacerbating malnutrition in these areas. Even in an abattoir that lacks modern facilities, such as the one in Ongata Rongai which supplies meat to Nairobi, a lot of meat from cattle, goats and sheep bought in drier areas is wasted annually due to cysts. In Lokichoggio, northwest of the Turkana district, there are as many as 200 cases per 100,000 population per annum. In Turkana, man is not always a dead-end intermediate host, since dogs and wild carnivores may have access to human bodies at times, due to inadequate burial of the bodies. This man-dog-man cycle for the parasite may enhance its pathogenicity for man and making it an important peculiarity of the disease in Kenya.

Sporadic cases of human hydatid disease were reported in other parts of Kenya by Okelo in 1981 and are increasingly being recognised. Despite these reports, there have been no national or country-wide surveys to determine the extent of the problem in all the provinces in Kenya.

During the last decade, the African Medical and Research Foundation (AMREF), through its Flying Doctor Service has been performing over 100 operations a year for hydatid disease. It is now realised that post-operative recurrence is a major problem in Kenya. These recurrent cases have severe cysts and are almost always inoperable.

This led to the search for medical treatment by chemotherapy, which was successfully introduced in the early 1980's, and has been a major contribution to the management of the disease. Kenya is also making useful contributions in research in various aspects of human hydatidosis such as the immunology of the disease.

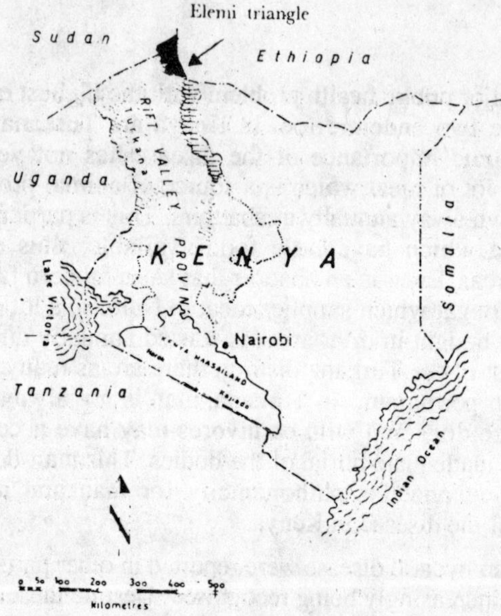

Fig. 1 Map of Kenya showing Turkana District, including Elimi Triangle and Masailand.

Fig. 2 Grave probably disturbed by carnivores.

Chapter 1

Occurrence of Hydatid Disease

Section I

HYDATID DISEASE IN TURKANA

Historical Aspects

Hydatid disease is not new as it was already recognised before the Hippocratic period. The disease has long been recognised as an occupational hazard in certain parts of the world, like in Lebanon where cobblers use dog faeces to tan leather. It is also a complication of remedies used in the past. An example of this is the use of dog faeces for treating rheumatism.

In Kenya, the disease was not recognised until 1915 when Leese first reported the larval stage in camels. However, the adult worm was not reported until 1958 when Ginsberg identified it from a jackal as *E. granulosus*. Following this finding, successive reports appeared and the disease became known as a major source of wastage of meat in abattoirs in Kenya, particularly in the central abattoir of the Kenya Meat Commission at Athi River. Hydatid disease has thus been appreciated as a major health problem in animals (cows, sheep, goats, etc) in Kenya.

The high prevalence of hydatid disease in Turkana district, a remote area, caused international interest which led to the arousal of interest in the disease throughout Kenya. It is therefore pertinent to look at the early studies by French, Chemtai and Okelo between 1976 and 1979 for historical interest. The new road to Lodwar and the vast development of the area in recent years has markedly changed the situation.

Review of the Situation Prior to August 1976

Turkana District in northwest Kenya, has been slow in developing into a modern world. After the discovery and naming of former Lake Rudolf (now Lake Turkana) by Graf von Samuel Teleki during his expedition of 1887-1888, the area to the west was visited by only a few intrepid explorers.

The early British administration regarded the Turkana as a nuisance and between 1903-1926 mounted many military expeditions to control them and maintain the area as a buffer zone between Kenya and Ethiopia. The first civil administration was established in 1926 after which Turkana was made part of Kenya.

The British administration did little to develop the area or to improve its contact with the outside world but used it for detention of those opposed to the administration instead. Around 1950 an interest began to be shown in the area and a number of surveys were commissioned by the government such as the one by P.H. Gulliver. An unpredicted series of floods and droughts between 1958 and 1961 devastated the area, causing the death of about 70% of the stock. The resultant famine forced government action and attracted missionary and other relief organizations, which until then had been tardy about taking an interest in this area.

It is therefore not surprising that before this period, there was practically no information concerning disease patterns or hydatid disease in particular. What did exist, though inaccurate, is anecdotal. For example, there is a verbal report given by a settler in Kitale, Mitford-Barbarton, of cysts having been seen during a journey to the area in 1910. A bibliography prepared for the Royal Geographical Society, South Turkana Expedition in 1969, had 333 entries, only one of which had some medical content. Much information for the period is probably hidden in official records or lost with them.

In the summary of available literature on hydatid disease in Turkana covering the period 1958-1977, reference is only made to the first known report and later work merely reproduces this particular information. An effort has been made to distinguish between data collected in Turkana and situations in which inferences have been drawn about Turkana, using data collected from elsewhere in East Africa.

In 1958, Wray was the first person to draw attention to the unusually high incidence of hydatid disease among the people of Turkana. Collecting data from various Medical Officers of Health for the years 1952 to 1955, he found 117 cases of hydatid disease of which 58 were in Turkana and 21 in Maasai, another pastoral people. This suggested an incidence in Turkana of 50 to 100 times more than in other parts of Kenya. It is important to emphasise that the disease occurs all over Kenya and indeed over most of the world with incidence in Turkana being much greater. Nelson and Rausch confirmed this high incidence and drew attention to morphological similarities of the parasite in Turkana with that in Alaska. They also collected data in animals to correlate it with that in humans.

It is interesting that in a special report on health in Turkana prepared for the Kenya Government in 1965, hydatid disease is not mentioned. The reports of intracranial hydatidosis by Clifford in 1968 and orbital hydatidosis by McClatche and Manku in 1967, while mentioning that the patients were Turkana did not pursue this fact, but were more concerned with the location of the cysts in the body.

Schwabe, using in-patient figures from Kitale and Lodwar hospitals, together with Gulliver's estimates of the population of Turkana, estimated the incidence at 40 cases per 100,000 people per annum. He also drew attention to the apparent lack of knowledge of the disease by the Turkana people themselves as evidenced by the lack of a Turkana name for the condition in humans. However, it is well recognised in animals where the cyst is known as *epespes*. The Turkanas now have this name for the disease in man.

Rottcher, in 1973, was the first person to publish results on a large number of patients he had personally treated while working as a surgeon with the Flying Doctor Service of the African Medical and Research Foundation (AMREF). Between 1968 and 1972 he carried out 1,990 operations of which 163 were for hydatid disease. Of the 163 cases, 142 were Turkana, the next largest group being Maasai with 17 cases. In addition to this information he reported that 97 were female and 64 male (a ratio of 1.5 : 1). Irving reported records from January 1964 to May 1970 for Lokori hospital which although situated in southern Turkana also draws patients from the northeast. There were 34 cases accounting for 4.5% of all the surgical operations carried out during this period at the hospital. After a two day visit to Turkana, Mann, quoting information provided by the Medical Officer of Health, drew attention to the fact that more cases come from the northeast and northwest corners of the district than elsewhere. O'Leary, the then Medical Officer of Health, who provided this information, later amplified it in a section devoted to hydatid disease in a book entitled *Health and Disease in Kenya*.

O'Leary later produced an extensive review of 789 cases treated in Turkana during the five years, 1971 to 1975. Nearly all the cases came from one of two areas; one extending from the northwest corner to the Pelekech mountain, the other in the northeast corner along the shores of Lake Turkana. As with Rottcher's observations, more females than males were affected and this was more noticeable in adults (female to male ratio; adults, 1.8 to 1; children

1.2 : 1). Over the years, between 1958 and 1977, information built up showing a very high incidence of hydatid disease in the Turkana people. The information indicates that females are more affected than males and that most cases come from the north of the district. All these data, however, have the disadvantage that they refer only to patients seeking treatment in a hospital. They can therefore only indicate the minimum rate of occurence of the disease, as some people do not seek hospital treatment. There was also lack of data on population distribution and movement as areas producing more cases of the disease might have had more people rather than have had a genuinely higher incidence.

That hydatid disease occurs in the area bordering Turkana has been demonstrated for northern Uganda, southern Sudan and southern Ethiopia. Hence it is the very high incidence in a small community which draws attention other than the community being unique in suffering from the disease. The transmission of the disease in this community may therefore have special characteristics. The presence of hydatid disease in animals and its mode of transmission to man in Turkana is, therefore, of great importance. Unfortunately, there is little information on this, and some of it was collected in other parts of East Africa and conclusions about Turkana drawn by inference.

The presence of hydatid cysts was noted by Ginsberg in animals slaughtered at Athi River, the central slaughter house for Kenya. At that time, animals coming from African owners had a higher prevalence than those from European-owned farms with different methods of stock rearing. The prevalence in African-owned stock in 1955 was given as 41.1% for cattle, 53.0% for sheep and 18.2% for goats. Similar observations were later made by Ginsberg and Froyd. These rates are for the whole of Kenya, but it seems reasonable to suppose animals in Turkana will have a similar or higher prevalence of the disease.

Irving reported in 1970 that among stock slaughtered at Lokori in the south of Turkana, the prevalence of taeniid cysts in goats was 59.4%. Only two sheep were killed and both were infected while the two cattle killed were not infected. He also drew attention to the scavenging by dogs in the slaughter area (Fig. 3) and the fact that cysts, if detected, were cut out of the offal and discarded to be eaten by these dogs. Sometimes they would be used as play things by the children before being made available to the dogs. This would account for the high prevalence of infections with a particularly large number of adult worms found in dogs in Turkana by Nelson and Rausch. In a study of dogs from all over Kenya, 27 out of 43 were infected. If only animals from Turkana were considered, a prevalence of infection of about 70% would be obtained against 50% for the rest of Kenya.

Several suggestions concerning the transmission of the disease have been made including the following:

i)   Transmission from the dog to domestic animals probably occurs in the stock yards at night when contact with dogs and their faeces is particularly close.

ii) Transmission to man is facilitated by the use of dogs to cleanse infants, allowing close contact with contaminated fur.

iii) The use of knives or cooking utensils contaminated when sharpening or cleaning with contaminated stones or soil.

iv) The use of contaminated clays and soil as body adornment.

v)   Use of contaminated plants or soils for medicinal purposes.

vi) Consumption of contaminated plants or berries as food.

vii) Contamination of milk during drying.

viii) Contamination of water due to sharing of water holes between dogs, humans and livestock.

The possibility that a sylvatic cycle might be of importance was raised by Wray when he suggested that humans might be infected by contact with jackals and hyenas. Hydatid cysts were found in 17% of buffalo and 10% of warthog in Uganda by Woodford and Sachs. Nelson *et al* found an infected wildebeest in Kenya. These species, however, do not live in the northeast of Turkana where the disease is common in humans. There is as yet no evidence that the wild herbivores in the area of Turkana where human hydatid disease is common are infected. Nelson and Rausch found adult worms in 3 out of 19 spotted hyenas, in 1 out of 9 black-backed jackals and in 3 out of 4 wild dogs killed in various parts of Kenya. Irving quotes Aldridge as stating that hyenas sometimes dug up human graves in Turkana and he questioned whether they and the domestic dogs might become infected in this way. This possibility was again mentioned by O'Leary and taken to dramatic proportions by Mann who envisaged whole livers being carried from human remains by birds of prey while hyenas devoured the corpses.

Fig. 3　　Slaughter area in a manyatta showing dogs stealing meat.

Early Hypotheses to Account for the High Prevalence in Man

Definition:　　　*Manyatta* is a whole homestead consisting of houses and stock yards.

Alar is a thorn fence surrounding a *manyatta*.

Akai is a house in a manyatta constructed using branches, leaves, grass, bark, etc. It is more or less hemispherical in shape measuring about 2 to 3 metres in diameter. The women and children sleep on skins in these houses.

In 1980 it became possible to re-examine the various hypotheses regarding the route of transmission of hydatid disease in Turkana. In doing this, a particular attempt was made to isolate the factor or factors present in Turkana that accounted for the exceptionally high incidence of the disease in humans in the district.

Several possible routes exist for transmission from the dogs to humans or livestock. As most of these are common to more than one host, the relative prevalence of the disease can be used to support or refute a particular suggestion. Especially interesting in this context are three observations made by French during his first three years in Turkana, and one by MacPherson and French, made shortly afterwards:

i)   The incidence of human infection varied in different parts of the district, from around 14 to over 200 per 100,000 population per annum.

ii) Adults had a higher incidence than children and females have 3 times the male incidence during child bearing years.

iii) The prevalence in most livestock was low. Goats, sheep and cattle had less than 5% while camels had a high prevalence, over 50%.

iv) Dogs in all parts of the district were infected with adult *Echinococcus* and the heaviest infections were in areas of low human incidence.

These various suggestions will now be considered in turn.

a)   Contamination of water holes

This is a distinct possibility (Fig. 4). There is a difference between areas like the Kerio, Weywe and Upper Turkwell river banks where the water is mobile and in areas where springs or water holes create small stagnant pools around which animals congregate. Here, faeces deposited are not washed away but remain to contaminate the water and the surrounding soil. This contributes to the spread of infections and is the most common case in areas of high human prevalence. Observations made at any of these water holes confirm that dogs have access to and defaecate in or near the water. However, as this is the primary means of spread, it is difficult to explain why humans in different areas have different prevalence rates even though the dogs are infected. It is also puzzling why females have greater chances of infection than males and why humans and camels are infected more than goats, sheep or cattle. It can be postulated that a particular strain of *E. granulosus* exists with a predilection for humans and camels. However this does not explain the male to female difference or the geographical differences as it is known that dogs are infected in all areas. The water hole is an obvious possibility, but fails to account on its own for several features of the distribution of the disease. Even though it is the most likely source of infection there must be other factors necessary to explain fully the present distribution of the disease.

b) Contamination of vegetation

The parts of the plants commonly consumed by humans are leaves of certain trees, berries, seeds and nuts all of which grow high above the ground and are unlikely to have come into contact with dog faeces.

Fig. 4   Waterhole showing a dog in close proximity.

A few roots which grow near rivers are eaten but this is in areas of low human prevalence. Similarly, the only other animal with a high prevalence of hydatid, the camel, browses rather than grazes. It consumes vegetation that is high above the ground and has been exposed to solar radiation of considerable intensity, so that anything on the surface is likely to have been dessicated. In the case of most leaves and roots consumed by humans a long cooking process is necessary to render them edible, usually 4 to 12 hours boiling and this will have rendered any eggs non-viable. As a result of these observations on the type of vegetation consumed and the methods of preparation, contaminated vegetation is unlikely to be the source of infection.

c) Contaminated soils

Ground temperatures in Turkana reach 60°C in the middle of the day and the soil is very dry except for a short time after rains. Eggs deposited would rapidly dessicate and become non-viable. Thus this route would not be able to sustain the high-prevalence rates. Also, this would by no means explain the large variations in incidence between areas in close proximity. Soil around water holes does contain enough moisture to support viable eggs. But this soil extends for only a short distance from the water and can be considered as being part of the water holes whose possible contamination has been discussed.

d) Contamination of knives and cooking utensils during cleaning

The substances used to clean these items are soil and water both of which have been discussed.

e) Drying of milk

This process involves placing a thin layer of milk on a skin and leaving it in the sun to dry. The hard crust obtained can be chewed or used in cooking. The skins used also serve as floor covers in the *akai* and therefore dogs will have slept on them and they may have been contaminated. Dogs also gain access to the milk during the drying process. However, it is necessary to carry the drying process to extreme lengths in order to preserve the milk. The fact that preservation occurs indicates that bacteria have been devitalized by the process. Although echinococcal eggs have greater powers of survival than most bacteria, it is probable that they will also have been devitalized. Another point against regarding this as the main route of transmission to the human, is that although this process is used in many areas of high prevalence, it is not common in some areas. It is even more difficult to account for the male to female difference since if there is any difference in consumption of the dried milk, then the male would be expected to have higher prevalence because the male eats more.

f) Transmission occurs in the stock yard

During the day, the animals either roam freely in search of vegetation or are herded to the water points. At night, they are collected inside protected areas surrounded by thorn fences. These areas are usually crowded and camels often have separate areas from the small stock. Most of the year, cattle are kept in another camp separate from the camels because of their different eating habits and the ability of the camel to survive in drier conditions. These fenced areas are called *manyattas* and include the human living accommodation. Dogs have access to all areas. These areas are dry and sand-covered with no vegetation or water.

It would therefore seem likely that if a dog defecated, the faeces would be consumed, and this would mean the herbivore actually lifting the faeces from the dry sand. However, it would not explain the high prevalence in camels compared to goats since dogs have equal access to all areas.

g) Close human-dog relationship allowing contact with contaminated fur

(A more detailed account follows under Turkana Customs in Section II)

The dog is traditionally part of the Turkana life. The men often have a pet dog which follows them and acts as a warning against intruders, either animal or human. However, many men will be seen without a dog in attendance. On the other hand, it is unusual for women to be seen without one or more dogs. These dogs have been specially trained to help care for infants. The dogs are characteristic and appear to breed true, being small (maximum 18 inches high), light brown in colour and naturally affectionate and nonaggressive. They sleep in the *akai*. This situation provides all that is needed for dog-to-human transmission. The dog is stroked and petted giving the possibility of contact with contaminated fur. The skins on which the family sleep can also be contaminated.

There are also possible explanations of the male-to-female and geographical variations. The female keep more dogs, particularly when child rearing, and so are more exposed than the male. The geographical variation may be due to the fact that the people in the south keep fewer dogs. There is approximately 1 dog to 4 people in the south, whereas the situation is reversed in the north, there being about 2 dogs to each person. Again, the amount of exposure may explain why the south which has fewer dogs and therefore less exposure, has a low incidence while the north with more dogs and therefore higher exposure, has a higher incidence of the disease (assuming identical source of infection in both regions).

h) Possible existence of a sylvatic cycle

At present, little is known of the prevalence of the disease in the wildlife of Turkana. That wild animals may be infected has already been mentioned.

MacPherson and Karstad examined a few wild animals in Turkana. No adult worms were found in hyenas, though they were found in jackals. Thus a sylvatic cycle is possible though it seems unlikely to be a significant source of human infection at present. In view of high prevalence rates in dogs and camels, there is sufficient domestic infection to account for the high prevalence of the disease in humans. Furthermore, if jackals or hyenas were a major source of human infection, there ought to be more males than females affected because it is the males who hunt and come into contact with the wild carnivores. Jackals and hyenas are hunted in Turkana.

Currently, too little is known of the prevalence of the disease in the wildlife in different parts of Turkana to make an assessment of the importance of the sylvatic cycle. Due to the close contact of humans with dogs, which are known to be infected, there is no need to postulate a wildlife source of human infection. However, a sylvatic cycle could assume more importance in perpetuating the disease during a control programme. This appears unlikely though and the probable source of infection for jackals is scavenging human settlements.

i) Burial customs and possible human to carnivore transmission

The methods of burial include exposure and shallow interment (Turkana Customs, Section II). Both allow possible access of domestic and wild carnivores to the corpses (Fig 2).

j) Medical use of dog faeces

Dog faeces is used as a dressing for cuts and septic sores. It is dried, mixed with berries, leaves and soil or charcoal, heated and after cooling applied to the wound. The heating process is sufficient to kill the eggs. So there is no danger of the recipient contracting the disease even if accidental transfer to the mouth occurs. However, the person involved in preparing the dressing is at risk. Only a small number of people are involved in the dressing and this does not explain the widespread distribution of the disease in Turkana. There could be a small group with an occupational risk.

k) Cleaning the *Manyatta*

The women clean the ground where the stock is kept with their hands and the dung collected is deposited in places outside. This is occasionally burnt. Dogs have free access to these areas and defecate in them. Therefore, during the cleaning procedure the hands will on occasions come into contact with dog faeces as well as dung. The eggs could be transferred to the mouth by the contaminated hands. This, being women's work, could place them at a higher risk than the men. It is unlikely that this is the only means of infection, but may supplement others and help explain the male to female difference.

Section II

DISCUSSION OF FACTORS INFLUENCING INFECTION OF THE DOG AND SPREAD OF HYDATID DISEASE IN TURKANA

Route of Infection of the Dog

It is surprising that while the routes of infection of the human have attracted so much attention, the mode by which the dog becomes infected has attracted very little comment. Irving reported high infection rates probably of *Taenia hydatigenia* in goats at Lokori, and noted that cysts were cut out of the meat and left for dogs to scavenge. This has been assumed sufficient evidence to explain the high incidence. Recent observations that do not support a high incidence of *E. granulosus* in goats, even in the Lokori area indicate that the camel is the most affected of the herbivores. No small wild animals which the dog might hunt have so far been found with cysts. Hence it must be concluded that the camel is the main source of infection for the dog. This is supported by the fact that practically all the camel cysts contain large numbers of viable protoscoleces. Even so, such a postulate raises another problem. Camels are rarely killed and thus access to the cysts must be limited. A probable explanation lies in the size of the camel and the occasions on which it is killed. Camels are killed on two occasions. One is when dying. In this situation the Turkana claim it is bad to eat the meat and offal. The other is during special feasts such as initiation ceremonies. In these situations there is likely to be an excess of offal. Meat and offal cannot be preserved for long and what therefore is left over is likely to be fed to pet dogs or left for stray dogs or wild carnivores to scavenge. Thus it is likely that if a camel is killed the dogs will be given access to, if not actually encouraged to eat infected offal. This situation could counterbalance the rarity of the event of killing a camel and maintain high prevalence of the disease in dogs.

Spread of Hydatid Disease in Turkana

The distribution of hydatid disease in livestock and therefore dogs throughout Turkana is probably maintained by the customs of stock exchange. The Turkana are great debaters with one topic being the begging or borrowing of stock. There is a continual small exchange of stock throughout the district.

Hypothetical Transmission Cycle : A Guide Developed in 1981 for Future Research

In conclusion, it was necessary to try to integrate the individual factors discussed into a composite cycle of transmission. The dog was thought most likely to acquire infection from camel offal. This is because the camel has a high prevalence of viable hydatid cysts containing numerous protoscoleces. Although camels are rarely killed, whenever they are, there is usually a surplus of meat and offal available to pet dogs and stray carnivores. Some of the offal made available to these animals is probably infected. The habit of large stock exchanges at weddings and other ceremonies might maintain the disease in all parts of Turkana district.

Transmission to livestock is most likely to occur at the water holes where conditions favour survival of ova in infective faeces. In addition, the water itself, or the small amount of heavily grazed vegetation around the hole can both be routes to the herbivores' intestine. Running water in streams is less likely to be a source of infection as the infected faeces is liable to be washed away.

The fact that the prevalence is high in camels, but low in other herbivores which have equal access to items contaminated by dog faeces, could be due to the strain of *E. granulosus* present in Turkana. It is necessary to postulate that the strain present is particularly adapted to camels and humans. The epidemiological evidence was in favour of this. Wild carnivores are infected probably by scavenging camel carcasses. At present there is no evidence that infection exists in wild herbivores but it may. The large amount of infection in domestic stock and lack of contact with wild carnivores suggests that the sylvatic disease is not the prime source of human infection. Even so the sylvatic cycle is of interest and could be of importance as a reservoir of infection when a control programme is undertaken. Human-to-carnivore transmission is possible, but would be a rarity and is unable to account for the high prevalence rate of human hydatid disease in Turkana.

Humans may be infected through water or using soil near the water holes for cleaning purposes. This however, is not in itself sufficient to account for the geographical, age and sex distribution of the disease. The serological tests for hydatid disease in humans in Turkana have a surprisingly high false negative rate. Okelo and Chemtai found that 40 to 50% of known cases of hydatid disease do not have positive responses in the standard serological tests. This could be due to the strain of *E. granulosus* or to an inherent trait in the Turkana themselves. If some Turkana are more susceptible to the disease than others, this might explain some of the differences between different areas, ages and sexes. While this possibility cannot as yet be refuted, a much more likely explanation lies in behavioural differences.

The risk of infection is probably related to the amount of human-dog contact. The number of dogs kept per person is higher in areas of high prevalence. Women during child rearing age require more dogs and this leads to greater contact with the dogs and could account for the higher prevalence of the disease in this group of women. The cleaning by hand of the manyatta is women's work and there is a risk of contaminating the hands with dog faeces. This could also explain the higher prevalence in women.

Human infection is most likely to be contracted by direct contact with infected dogs. Water or food borne infection may occur on a few occasions, but in itself does not explain the distribution of the disease. That infection is more likely the greater the contact with the dog, can explain the known features of the distribution of human hydatidosis in Turkana.

This postulated cycle indicates a lack of detailed knowledge of dog feeding, amount of human-dog contact and strain of *E. granulosus* in Turkana. These, therefore are the areas on which future research should be focussed. The postulated spread by human-pet-dog contact offers a possible intervention for control. If it is the pet dog that is the primary source of human infection, dosing with an anti-helminthic should have a dramatic effect. Preliminary studies indicated that owners would accept dog treatment after appropriate explanation.

Turkana Customs

It is clear that behaviour plays an important role in the spread of hydatid disease. Therefore observations were made on this subject in Turkana. Although reported in the annual reports of AMREF, some lack of continuity has been noted. Therefore, this attempt has been made to edit and amplify these reports into a single entity.

a) The Turkana

The Turkana are an individualistic, hardy people living in the inhospitable northwest of Kenya (Fig. 5). The area is known as Turkana District, but the Turkana also occupy an area to the northeast known as the Elimi triangle.

The area is amazingly variable from hot water falls and near boiling springs around Kapedo in the south, the sandy Nachorugwai desert on the mid-west bank of Lake Turkana, the swamps of the Loikipi plain in the north to near tropical jungle on the Puch central highlands over 1200 metres above seal level. More typical are the dry plains clotted with accacia thorn trees and savannah plants, that lie between mountains covered with scanty grass and scrub. In the west, rivers rise in the mountains and flow down into these valleys before drying up. In the dry season wells up to 6 metres deep have to be dug in the dried-up rivers in the valleys and plains. Around the mountain ranges, a few permanent springs exist. The Turkana's life is thus a continual search for water and pasture for his stock. Often there is pasture , but no water or vice-versa.

The Turkana originated from the plains Nilotes. These moved into Eastern Africa and split into two groups; one moving south to form the Masai group, the other remaining in the northwest forming the Karomojong group.

Fig. 5   Turkana Group.

After a violent power struggle, the Karomojong group split into Karomojong and Jie. Due to shortage of grazing land some Jie moved down the west escarpment of the Great Rift Valley and became the Turkana. Both tribes have legends concerning this friendly split.

The Turkana have no particular law or ruler, rather each man is head of his own household. Brothers, sisters and half-brothers may retain friendly ties forming a family group, but each household is independent and proud of this fact. The clan groups seem to have originated from family groups. Members of a clan can usually find they have relatives in common, even if somewhat distant. The clan has no power other than friendship and similar interests in an area. However, members have features of dress and behaviour in common. Examples can be quoted such as the *Ngimataperi* who are reputedly aggressive and prepared to indulge in inter-tribal fighting, the *Ngitobotok* and *Ngiboicheros* who live on the banks of the Turkwell River and have developed a form of agriculture and the *Ngisonyoka* many of whom, for unknown reasons, do not drink blood, unlike the other two groups.

Another division is the alternation of men into leopard or stone groups. A male child takes the alternation opposite to his father if legitimate, and takes that of his mother's grandfather if illegitimate. This division has ritual significance. A young man must find a man of his own alternation to sponsor his initiation. Also at marriage the groom gives the bride a necklace made of a single strand of thick wire. In the case of a member of the leopard group this is coloured copper while in the case of a stone group it is coloured tin.

Since 1970 various developments have occurred and made an impression on these traditional divisions. At present the population can be divided into four groups:

i)   Traditional, living along the bases of mountain ranges near dry river beds.

ii)   Fishing communities along the lake shore mainly established during the 1970's through the efforts of NORAD aid programme.

iii)   New settlemnets at minor irrigation schemes, started in the 1960's and being extended in size and number.

iv)   Developing urban communities around larger towns, at present mainly centred around Lodwar, with smaller groups around Lokitong and Lokichoggio.

A further division occurs throughout the Turkana community and sub-divides each of these four divisions into two groups. This is between those who possess livestock and property, and a "beggar" class called *Masakeenem*. All groups like to keep on the move, but in the stock owning group movement is more organized and dependent on the needs of the animals.

b)   Movements

The Turkana are a nomadic people and movement plays a major part in their lives. It can be a significant factor in the spread of disease. The following observations were made during the early part of this study.

The desire to travel extends to all age groups even to primary school children, who frequently ask for lifts in vehicles. They rarely ask to go to a specific place, but if pressed they ask where the vehicle is going and then announce with delight that that is just where they wanted to go. They appear to have no fear of survival in strange places or of the likelihood of a very long walk home.

Movements that have been observed have been classified as follows:

i)   Regular cycles either daily or every few days. These involve taking the animals from the *manyatta* to water or graze. Quite often *manyattas* are sited between water in a valley and grazing land on a hillside. The *manyattas* are passed during movement between these two places. Where the distance between the water and grazing land is short, the *manyatta* may be situated at either, regular journeys being made to the other. As has been stated however, surface water is avoided probably because of the mosquitoes and other pests which abound near these sites particularly at night. This normal reluctance to live near surface water creates problems for fishing communities and irrigation workers.

ii)   Shorter regular cycles each day to fetch water and occasionally food to the *manyatta*.

iii) Inter-*manyatta* movements usually by the men, particularly the headmen. The Turkana men are allowed to have as many wives as they can afford the dowry for, and the wealthier the man, the more wives he has. The largest number known is 58 still living, but more usually it is two or three. It is common practice to set up *manyattas* in more than one area, and a wife, younger brother or grown-up son is needed in each. This is necessary to cater for the differing needs of the different types of stock. Camels and goats tolerate dry conditions and browse, while cattle and sheep need more water and grazing. This is also an insurance against a natural disaster in one area. The owners therefore have to travel between the *manyattas* regularly to visit their wives and children and supervise the stock rearing.

      In smaller families independent brothers or friends may agree to cater for different stock needs, as an insurance against natural setback. This also requires movement between *manyatta* for liaison.

iv)   Trade, including payment of dowry involves moving stock from one area to another.

v)   Movement of the whole *manyatta*. *Manyattas* are moved at variable times and for varying reasons:

      (a)   *No movement.* An extreme case is known where a *manyatta* is still on the site selected by its present owner's grandfather (approximately 18 years). In fishing communities and around irrigation schemes there is a tendency for *manyattas* to become stationary.

(b) ***Short movement.*** This may be only a few yards and the reasons given have been the appearance of insect pests, flooding (rains cause a large number of cases) and bereavement, particularly of a married woman with children, when the custom is to collapse her *akai* over her on her death and leave the *manyatta*.

(c) ***Longer movement.*** This is in search of grazing land or water. It is quite often cyclic with the same family returning to the same area at a particular time each year and involves occupation of two to four sites in order during the year. The high ground is occupied during dry seasons and the low when wet.

(d) ***Spontaneous long movement.*** Three cases are known where the family and stock moved from one area to another. The reason given in each case was a vague hope for better conditions or a quarrel with neighbours.

The movement from one area to another, particularly when long distances are involved, is through well defined routes. These are clearly visible both on the ground and from the air, and have been extensively used for many years. Haphazard wandering occurs occasionally when the owners get the urge just to move. Three of the groups mentioned in the introduction have made a more or less permanent settlement. These are the fishing community, the irrigation scheme workers and the urban groups around large towns.

Voluntary fishing is attractive due to the high monetary rewards from fishing. These rewards were so great that in the late 1970's they attracted fishermen from as far away as Lake Victoria who came to compete with the Turkana. These communities are relatively wealthy and use their wealth to increase stock. They still have the urge to return to the traditional lifestyle, fishing being a way to increase wealth. They have adopted many western customs and use profits from fish trading to buy radios, torches, etc. Some still wear traditional clothing, but many have adopted western dress. During seasons when mosquitoes are a menace, most people move away from the lake shore. The Turkana Fishermen's Cooperative Society and several independent traders offer them a ready market for their fish. Since food is plentiful, both from the fish and the large herds the family keeps, much offal is discarded. In the fish cleaning areas, large numbers of dogs scavenge for food, and in these areas there is very close contact between man and dog. Since the dogs eat fish, they do not get infested with *E. granulosus* and so pose no threat of spreading hydatid disease. The urban group is mainly composed of artisans and labourers, but there is a rapidly growing number of educated Turkana in the administration, police and other government and non-government organisations. Again there is a reluctance to give up their traditional way of life and many families get split, with one brother, usually the eldest, remaining traditional, while another goes for education and into urban life. This is regarded as a further analogous situation to the division of a family and its stock between several areas. Should one fail, the others are likely to succeed and make good the loss. This urban group, particularly the educated, have adopted many modern concepts of hygiene which ought to reduce the risk of infection. However, their frequent visits to traditional relatives or to their own traditional *manyatta* probably places them at equal risk with other Turkana to infection with the common diseases.

The group working on the irrigation schemes is very different being drawn largely from the impoverished and destitute. The famine of 1959-1960 and 1980 - 1982 left many Turkana without stock. When the famine camps were closed they had no means of support except to take work on the irrigation schemes. Pay is low, often given as maize meal grown on the schemes. These people have little chance to improve their conditions and continue to live in the traditional way. The location of the schemes near surface water and stagnation of water in the irrigation channels both produce vast numbers of mosquitoes and other insect pests. The Turkana attempt to ward these off by lighting fires which are then damped down to produce smoke. Malaria is a very common parasitic disease. A government survey at one site in 1976 found parasites of one form or another in over 30% of a random sample of workers, none of whom was complaining of any illness. The low hygiene and traditional way of life would seem to place these people at a higher risk of infection than other Turkana. There is no evidence at present to indicate that settlement has changed the incidence of hydatidosis. This may be in part due to the frequent visits to traditional areas and in part due to the incubation period being longer than the period during which settlement has taken place.

c) Special Groups of Turkana

i) The *imuron*

These are traditional medical practitioners who are usually wise old women. These groups overlap in their activities and one person may, in effect carry out functions in more than one section.

The *imuron* is a fortune teller with additional powers to heal and to prevent evil. The future is usually foretold by throwing sandals and then divining the future from the way they land. This practice can be further extended to help in decision making. For example telling the future for two or more different actions so that the most beneficial can be selected. Charms can be used to protect against evil spirits or offerings made to good spirits or to the high god *Akuju*. The traditional medical practitioner also uses fortune telling to determine the progress and treatment of a disease. Burning the skin and drawing teeth are common cures. Herbal remedies are made, more usually by the wise old women. A dressing containing dog faeces is used. Since it is baked, the dressing is unlikely to contain viable *Echinococcus* eggs. The wise old women make up herbal remedies and the dressings mentioned earlier. In addition they have their most important role, which is that of midwives.

ii) *The elders*

These are old men respected for their knowledge. They are consulted on matters of ethics, customs and religion as well as subjects for which they are reputed to have a special expertise. They carry out ritual functions, but have no power to enforce their views. Each man is free to take or not take the advice.

iii) *The ngoroko*

Mr. Arnold Hopf provided information on the *ngoroko*. The following is a summary of that information: At the beginning of the second world war, the British thought an invasion of Kenya by the Italians might occur through the nothern territories including Turkana. A Turkana defence force was therefore organised and issued with ex-First World War rifles and ammunition which only occasionally detonated. When Abyssinia (Ethiopia) fell, the threat of invasion ended. The Turkana force was never properly disbanded or disarmed. They started stock raiding southwards. Hence, the name *ngoroko* which means "Men from the north" in Turkana language. Since the 1940's, any bandit has had the term *ngoroko* attached to him. The numbers have grown and in particular there is the ex-Ugandan Army element, most of whom have modern weapons. They pose a considerable threat to the traditional Turkana who are armed with only a spear. Thus the areas where they operate are largely deserted.

d) Comparison of South with North Turkana

The south of Turkana is now practically unpopulated due to the activities of the *ngoroko*. Strenous efforts are being made by the Kenyan authorities to control the security situation. A small number of stalwart Turkana remain with a few animals in the area to preserve their traditional grazing rights. Moving north, it is not until Lokori is reached that substantial numbers of Turkana are found. These upheavals make comparison of habits difficult at present. Two factors are, however, noticeable in the south:

i) The hair-style: The man's chignon is larger, involves the whole head, and is rolled across the head. The women shave less of the head and hence have many more plaits of hair covering nearly the whole scalp.

ii) The *akai* are simpler, with the bark being the main covering often so incomplete as to be seen through.

Some houses are raised off the ground on stilts, and in some cases the day house is built directly underneath. The entrance is much less restricted and strangers are welcomed inside. Further north, the covering is of grass thatch, or palm leaves and is thick and complete. The entrances are covered with skins and entry is restricted to family members. A simpler day house to which married women with children are entitled is built alongside for entertainment. In the more strict families even doctors are prevented from entering. The patient is brought outside by the women. The large and rapid enforced movement of the population from the south may have had the effect of distorting the data and thus no firm conclusions can be reached at present.

Water is more readily available in the south, which means that it is not so necessary to have dogs clean the infants by licking. The number of dogs in each *manyatta* in the south has been confirmed statistically to be less than those in the north.

e) The Relationship Between Man and Dogs

The relationship between man and dog has been assumed to be the main source of infection. Therefore this has been observed at every opportunity and is reviewed here in detail.

Many Turkana families possess dogs, which are trained to some extent. Their main use seems to be to eat offensive matter. In the small enclosed space where water is used for survival and not for cleaning, it is often observed that a woman who wants a mess cleared up will tap it with a stick, whereupon a dog will come and eat it. No apparent reward is given to the dog for this service.

The practice of allowing dogs to lick infants (Fig.6) after vomiting or defecation is seen, and may contribute to the spread of hydatidosis. The lick itself may transfer ova if the dog has previously licked its own anus. It is also an indication that dogs and man are closely interlinked. As previously stated this was necessary in an area so short of water that none can be "wasted" on washing. Many had to carry their drinking water for over 2 miles from the nearest water hole. An additional function has come to light. The dog protects young children from attacks by wild animals such as jackals. These dogs are valued and many owners are unwilling to part with them even for money. When they agree to have their dogs killed, they request that this should be done in their absence. During this period the child will naturally develop an affection for the dog. As the child grows up it may be given or adopt a particular dog, giving it small scraps. This relationship tends to end in the teens for men, but continues for women, who during child rearing may require several dogs. On the other hand the male during adolescence is given a particular ox and it is on this that his main affection lies. This animal should never die naturally, but is slaughtered by his friends when it is ageing or sick, and then replaced with another one. A man may still use dogs to give warning of predators near his stock, but many men are not continuously followed by a particular pet dog. The sight of a woman with an infant having a dog in close constant attendance was almost invariable. These specially trained dogs are regarded with great affection. They attach themselves to the women, by whom they are occassionally fed to retain their services as cleansing agents. The men normally chase their dogs away. This predilection may explain the higher incidence of the disease in women particularly during child rearing period when the dog is more essential.

f) Contact of Humans with Dog Faeces

The closeness of the contact of man and dog in the *manyatta* and *akai* is illustrated by an incident observed in a *manyatta* near Kibich on Thursday, 19 January 1978, which will now be described. The *manyatta* was a large one consisting of between 500 to 600 people living inside an outer *alar* about 150 metres by 50 metres, each family having their own *akai* (Fig.7). The people were of the Dongora division of the Turkana and unused to medical visits. One was made on this occasion because of an outbreak of influenza with a high child mortality. One house was about 3 metres in diameter and 2.5 metres high, the only opening being the entrance which was covered with skins, as was the floor. A small fire was burning, filling the building with smoke and there was the pungent smell of old camel's milk. Two adult women were present, the older breast-feeding an infant, the other sitting by two very ill children. In between these two women were two small goats tied to the wall. These, it was learnt later, were twins, the result of a premature delivery the day before. Next to them was a bitch with four puppies. While the ill children were being examined, the older woman placed her infant on the floor next to these animals. At one stage one of the puppies was seen licking the baby's anus, a clear way to obtain a

Fig. 6   Dog cleaning an infant.

massive dose of ova, had the dog been infected. The human infant was in a position to have obtained just as close a contact either by hand or mouth. The dog was used to suckle the goats as they could not suckle from their mother. They were therefore being dog and hand fed. The proximity of the animals was deliberate, probably also for warmth. The explanation given was vague, indicating that it had been empirically determined that this was a good way to save goats in this situation. The close proximity of infant and adult humans to dogs and livestock in the *akai* is reason for which infection with large quantities of eggs is likely.

g) Contact of Dogs with Animal Carcasses

When a small animal is slaughtered numerous dogs gather around, but are driven off, only being allowed to lick the blood-stained soil. In one method of slaughter the goat or other small animal is pierced from the rear under the ribs. A small hole in the abdomen is made and the guts removed for cleaning.

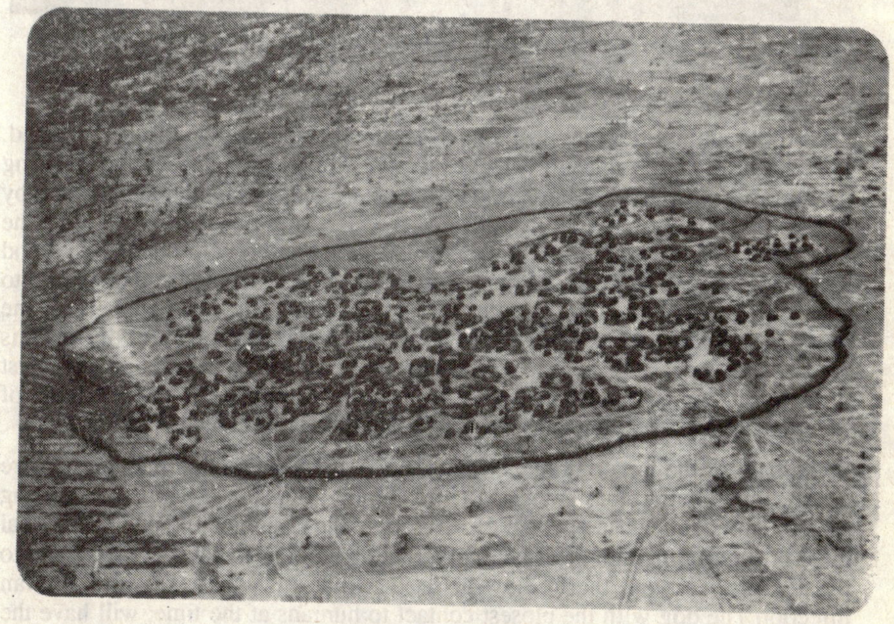

Fig. 7 (i)   Aerial view of a large manyatta.

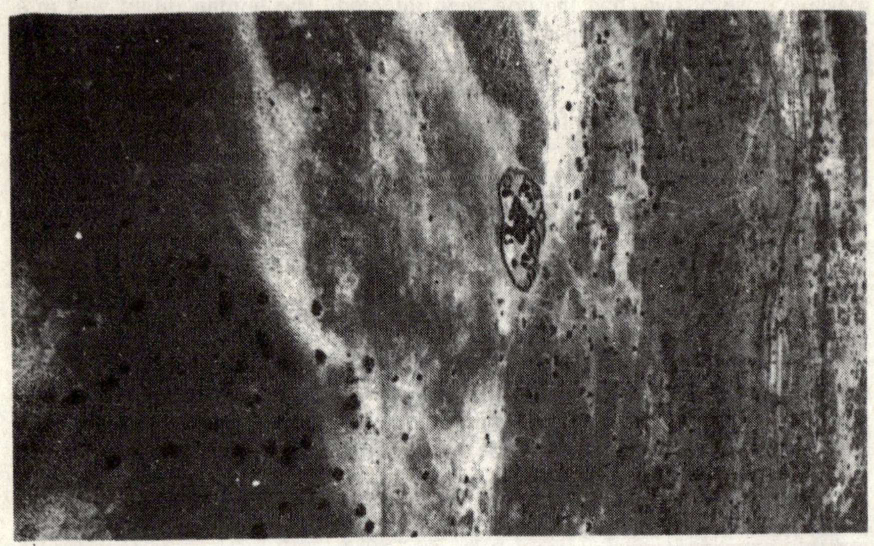

Fig. 7 (ii) Aerial view of a small (family) manyatta.

The whole animal, including the guts, is then placed on the fire and roasted. There is little opportunity to detect cysts, which will probably collapse during cooking and be eaten by humans. The bones are pulverised and again eaten by humans. Any item discarded is collected by *Maskini*, none being offered to the dogs, which are chased away. In the other method the throat is cut and the blood collected in bowls for later consumption. The animal is skinned and divided into many joints, each having a certain traditional value. These are cooked for the person entitled to them when desired. Cysts may be discovered during this process and are usually burnt. Dogs may lick the soil contaminated with cyst fluid or eat the cyst wall and thus become infected. These two methods of slaughter are extremes and a specific case may involve elements of both .

In the case of the larger animals, camels and cattle, the second method is more usual as there is rarely a gathering large enough to justify a whole beast being roasted at one time. The pet dogs are present at the slaughter and attempt to steal meat or blood. Infected offal including that containing hydatid cyst is given to them. In this way there may arise a situation which favours human infection.The dog with the closest contact to humans at the time, will have the greatest chances to steal an infected offal, Fig.3.

i) Feral dogs

With their domestic dogs, as with their livestock, the Turkana make no effort to limit the offspring. This results in a rapid increase in numbers and many dogs are driven out. These roam *manyattas* and settlements in packs scavenging for food. They gradually increase in numbers until their attacks on young livestock, particularly goats, become a menace. At this stage hunting parties, usually of boys and young men are sent to hunt them. The killing is usually by throwing sticks and is a prolonged affair during which the dogs frequently defecate, contaminating the pasture and occassionally the hunters.

ii) Wild dogs

The true wild dog, *epiot*, is an entirely different animal. It is larger and more aggressive than its domestic counterpart. It is rarely seen except when it is hunted because of its threat to livestock. The only ones seen during this survey were in south Turkana.

They do not play a part in the infection of man and his livestock, because of their rarity.

h) Burial Customs and Possible Human-to-Carnivore Transmission

The disposal of the dead varies from the elaborate ceremonies on the death of the head of a household, with complete internment, followed by a second series of ceremonies about six months later when the stock is divided between his heirs, to merely leaving that body or dying person where he/she fell. Much depends on the status of the dead person, the clan customs and the circumstances at the time. In the case of people dying while herding stock in remote areas, or as a result of violent clashes, there is often little choice, but to leave the bodies. In the case of infants and young children the body is left in the shrub near the place of death, often with a simple but moving ceremony. The mother first indicates that she loved the child then turns east towards the place where the sun rises and by throwing stones or other gestures, indicates to *Akuju* (god) that it is behind her and she wants more children. Women who have not borne children are removed from their *akai* through a hole in the back wall and placed outside the homestead. A married woman who has borne children merits having the *akai* collapsed over her and the homestead is moved to a nearby site or if this is not convenient she is buried outside the homestead. Most adult men who have acquired stock, wives and families will be given the treatment mentioned at the beginning of this section. Although this is the situation usually reported by Turkana, there are regional variations. Details are difficult to obtain because there are practically no reports from external observers who were actually present at such proceedings. Only one person, Hopf, is known by us to have actually witnessed the hut collapsing ceremony.

What is important from the point of view of the spread of hydatid disease is that carnivores, including the domestic dogs, may have access to human remains (Fig.8), and that these might contain viable hydatid cysts. There is as yet no estimate of the proportion of hydatid disease in buried humans, against those left exposed. Observations made on animal carcasses indicate that even large bones and skulls are rapidly scavenged, usually in under 24 hours and at night. This is the time when the greater proportion of domestic dogs are restrained in the homestead. So most of the scavenging must be done by wild animals or feral dogs. Thus, it appears that any scavenging of human remains is done by wild carnivores or feral dogs and the bodies are more likely to be young with a lower prevalence for cysts.

From the incidence of the disease, it has been calculated that no more than 100 persons per year died with hydatid cysts in an area over 65,000 sq km. This is probably too small to maintain infection rates in domestic dogs. Also the domestic dogs are infected in areas with low rates of human prevalence. The situation in wild carnivores in these areas in not yet known.

Briefly, it appears that though human to carnivore transmission may occur this is of interest as a rarity and cannot account for the very high incidence of hydatid disease in Turkana district; it may lead to selection of strains of *E. granulosus* particularly virulent in humans.

i) Relationship Between Man and Livestock

The animals are branded, often with very intricate, artistic designs to aid identification. Most, however, are known personally and have been given names by their owners. The names are usually descriptive, depending on colour, size and some other feature of the beast. One old man when demonstrating his herd of some 22 cows and 96 goats was able to name each. He introduced his favourite beast, a cow, but even the interpreter, himself a Turkana, failed to grasp the complex name. After about half an hour of diagrams in the sand, it became clear that the name depended on a joke about the right horn being normal and the left being twisted and pointing downwards. This was reminiscent of "the cow with the crumpled horn" of English nursery rhyme fame. This is now the beast's name.

The animals are considered to aid their owners in more ways than just providing food. No mention had been made of this until a patient who had been given ketamine anaethesia in order to suture multiple cuts received in a knife fight, regained consciousness in a happy confused state, entertaining the ward with songs until a sedative took effect. When the song was interpreted, it was a traditional bawdy ballad, calling on the beast, his best bull, to come and give him strength through male power, in order to overcome his opponents.

Fig. 8   Human remains near a grave probably opened by carnivores.

Care of the livestock is taught from birth and forms an essential part of the Turkana character. On one occasion a five year old girl was observed for nearly two hours trying to get a young goat to suckle. If she let go the young goat to catch the mother, it wandered off, but if she held it in her arms its weight impeded her movements to the extent of being unable to catch the mother. Eventually she succeeded in approximating the animal and gently milked the mother into the young animal's mouth before retiring completely exhaused to rest under a tree.

On another occasion a young boy of the same age was seen catching a goat and taking the milk directly into his mouth. This provides an explanation as to how these young children can survive a day in the hot sun without much water. It might also provide a source of infection if the goat's udder was contaminated with dog faeces, but this is unlikely.

j) Stock Exchanges

The main source of infection for the dogs appears to be from the domestic livestock. Therefore, the movement of livestock is important in the spread of the disease in the district. The Turkana are inveterate borrowers. In fact they will often attempt to borrow an animal for no other purpose than the entertainment provided by the protracted arguments this involves. All aspects of social life are connected with stock including birth, initiation, marriage and death, which are all celebrated with feasts of meat. This usually involves the borrowing for use of animals with an understood obligation to return these in due course.

Since marriage involves the largest single exchange it will be considered as an example. The present bride price is 30 to 50 head of large stock, camels or cattle, and up to 500 head of small stock, goats or sheep (equivalent to about US $5,000). Therefore, considerable movement of stock is involved. The groom has to raise this by borrowing from every possible source, such as relatives and friends. In doing so he collects stock from widely scattered parts of Turkana. Once this is given to the bride's father, he distributes it equally and widely, either to repay old debts or to gain favour so that loans can be obtained in the future. In this way, stock is mixed from widely separated areas. The explanations given as to why events are celebrated in this way is that *Akuju* will punish the offenders by injuring their herds. This may be a traditional explanation of new stock improving the breed of small herds. Unfortunately, it also ensures that any disease in livestock, particularly a long term one like hydatid disease, is spread throughout Turkana.

k) Diet

The main item of diet is milk, usually from the camel. Goat's milk is considered particularly suitable for infants and young children because the richer camel's milk, with its high calcium content, might cause digestive disorders. This habit is probably the result of experience over many years.

The milk is drunk fresh or slightly sour after about 24 hours. If it is to be stored longer, it is converted to a butter cream by shaking in the milk bottles made of goat skin. One or two milk bottles are suspended from a beam or part of the house structure and swung backwards and forwards by one of the women until the process is complete. The liquid is then ready for cooking or drinking. In some areas, particularly in the northwest, the milk is dried in the sun on skins and the solid matter stored. This process involves placing a thin layer of milk on a skin and leaving it in the sun to dry. The hard crust obtained can be chewed or used in cooking.

1) Early Hypotheses to Account for the High Prevalence

i) Blood from camels, cows and goats is drunk, but probably does not provide a major source of energy. The blood may be taken alone or, more usually, mixed with milk to flavour.

ii) Meat is usually grilled over embers, and occasionally boiled. In towns it forms an important source of nutrition, but it is less commonly eaten in the countryside where animals are preserved for milk.

iii) Fish is boiled, but is available only along the lake shore and in the larger rivers.

iv) Leaves are collected and cooked in all areas. These include creepers, e.g. *Edapal* requiring only two hours boiling. Bushes e.g. *Edung* and trees, e.g. *Ekale* need over twelve hours cooking. Rhizomes e.g. *Akali* are collected in the river beds and also boiled. Since these items are cooked, it is unlikely that they form a source of infection.

v) The seeds of acacia and other trees are collected and hand crushed between stones. The powder "flour" is used to make a form of bread or cake after mixing with water and cooking in the sand under a fire.

vi) Millet, maize and other crops are cultivated near rivers in small areas enclosed by thorn hedges. The seeds are ground and the flour used for cooking.

In these areas, patches of legumes may be similarly cultivated but are usually used to fatten stock before slaughter. The nuts of the daum palm have a hard, bitter, fibrous layer which is very popular. It is chewed raw, and stimulates salivation, thus avoiding the dry mouth experienced in this hot and dry area. It is occasionally soaked in the blood and milk mixture to flavour. In the main markets such as Lodwar, many stalls sell this delicacy.

Section III

HYDATID DISEASE IN MASAILAND

Historical Aspects

Although the disease has existed in this area for centuries it attracted little attention. Nelson and Rausch in 1963 reported the finding of adult worms in jackals, hyenas and wild hunting dogs from northern Masailand. Rottcher in 1973 reported 17 human cases in Masai but, was unimpressed by this, because of the very much larger number of 142 in Turkana.

The only systematic study in this area is that of Eugster during the period 1969-1976. Unfortunately, after using this data for his PhD thesis in 1978, he did not publish it more widely. The most important finding is that unlike Turkana, there is a wildlife cycle, as well as a domestic one. These two cycles seem to be independent and to have little effect on each other.

Sylvatic Cycle

The main definitive host appears to be the jackal, both from the number and prevalence of infection, 38%. The wild hunting dog and hyena have also been found to be infected as has the lion, but these probably play only a minor role in the maintenance of the cycle. MacPherson and Karstad have added a further factor suggesting that the hyena cannot play a significant role. They suggest that it is relatively resistant to artificial infection compared to the dog or jackal. The main intermediate host is the wildebeest, 12% infected, but cysts were also found in Grant's gazelle, blue duiker, impala, kongoni and buffalo. Although game animals are found all over Masailand, the main concentrations are in the National Parks, Amboseli to the east and Masai Mara to the west. Contact between humans and these animals is limited, particularly since hunting is at present banned. Thus, although of scientific interest this sylvatic cycle plays only a minor, if any, role in the causation of human disease.

Section IV

DIFFERENCES BETWEEN MASAI AND TURKANA

Several major differences emerge when the work of Eugster in Masailand is compared to that of French in Turkana. This is in spite of the fact that the terrain, vegetation and life styles are very similar. There is more rain and therefore more available water in most parts of Masailand. In both areas, as far as human disease is concerned, the sylvatic element may be ignored, thus only the domestic cycles are considered here.

Relationship with the Dog

In Masailand this is much less intimate because the presence of water means that the use of the dog as a nurse to cleanse infants is not necessary. It may be done in some of the drier areas. Another noticeable difference is that the dog is not allowed into the dwellings. In this context it is interesting that Cox reports the same for the Pokot who although neighbours of the Turkana have a very low prevalence of hydatid disease.

Infection Pressure

Probably the most important difference as far as the dog is concerned is the intensity of infection. Eugster regarded 20 worms as a heavy infection. In Turkana this might even have gone unnoticed since French and MacPherson were looking at specimens containing hundreds to thousands of worms.

In the end the often repeated rule applies, *the more the dogs, the greater their prevalence of infection and the heavier the infection the greater the infection pressure and the greater the infection rate in intermediate hosts* including man.

Intermediate Hosts

The Masai do not keep camels. In their area, cattle seem to be the main intermediate hosts with around 47% having cysts. Sheep with 30% probably also make a significant contribution but goats with only 9% are unlikely to do so.

It therefore appears that the intermediate hosts in Masailand's domestic cycle are cattle and sheep, while in Turkana they are the camel and goat. This difference is due to the drier conditions in Turkana favouring the hardier animals with a tolerance for water deprivation. However, it does raise the possibility of a strain difference and hence human infectivity by *E. granulosus* between the Turkana and Masailand. The work of MacPherson, while supporting strain differences, does not indicate that those in Masailand and Turkana are different. Hence, it is behavioural differences that are the major cause of the differences in human prevalence of this disease.

Chapter 2

Life Cycle and Epidemiology

Section I

LIFE CYCLE

The domestic cycle (Fig. 9) is the one that affects man. In Turkana district it is the only cycle, while in Masailand there is a wildlife cycle, but it is unlikely that this affects man. The domestic lifecycle consists of the domestic dog as the definite host harbouring the adult worm (*E. granulosus*) while the domestic herbivores, camels, cattle, sheep, goats and donkeys and man constitute the intermediate hosts. The adult worm in the dog lays eggs (Fig. 10) which come out with dog faeces, contaminating the grass and water holes. The intermediate hosts become infected during grazing or drinking. The ingested eggs form cysts (Fig. 11) (the larval stage of the parasite) in the viscera and other tissues of these intermediate hosts. Thus, when these animals are slaughtered and their raw infected offal is fed to dogs or scavenged by other carnivores such as jackals, the latter get infected. The "Worm-heads" of the protoscoleces evaginate (Fig. 12) and the larva grows into an adult worm (Fig. 13) in the dog or jackal. One dog may have from one to hundreds of thousands of these worms measuring 3-8 mm in size. On inspection of the intestines of the dog one sees them literally "carpetted" by millions of these worms (Fig. 14a).

Man is an accidental, usually dead end, intermediate host. Humans get infected by ingesting *Echinococcus* eggs from hands contaminated when handling dogs, or eating or drinking contaminated food or water. The eggs get embryonated through the action of bile and the resulting oncospheres hatch producing embryos that enter the portal circulation to reach the liver (the first filter), the lungs, brain, etc. Thus, the eggs can reach practically any organ in the body via blood.

Where dead human bodies have occasionally been accessible to dogs and jackals, thus facilitating the man-dog-man cycle, it is thought that this enhances the pathogenicity of the parasite.

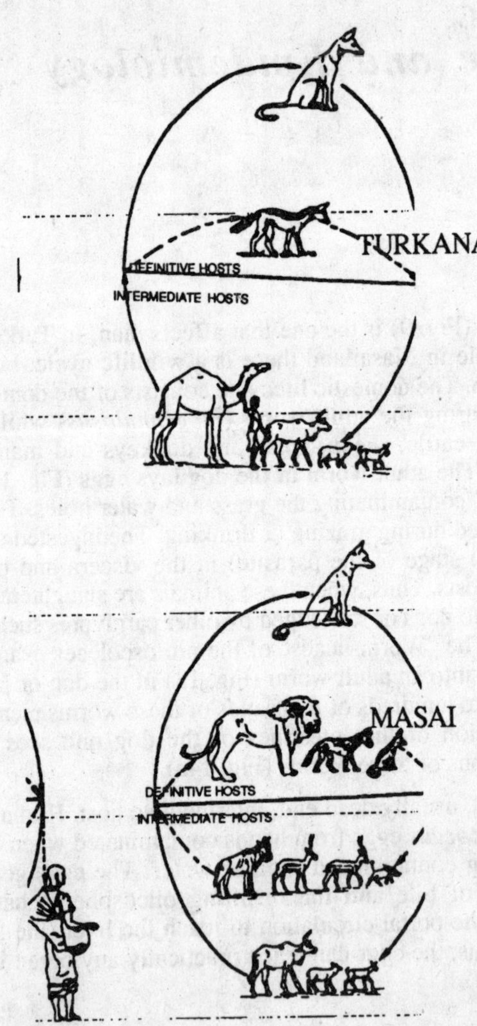

Fig. 9 Life cycle of *Echinococcus granulosus* in Turkana and Masai.

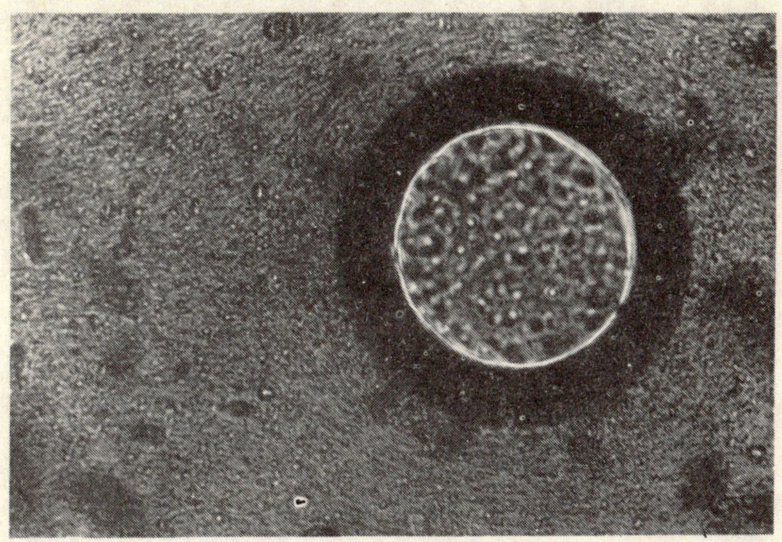

Fig. 10 Taeniid egg from source known to be that of *Echinococcus granulosus*.

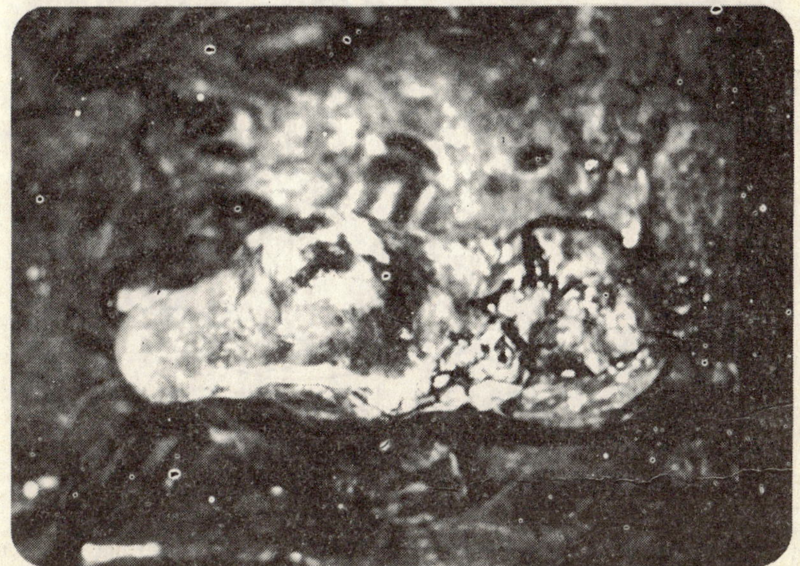

Fig. 11 Camel lung cyst.

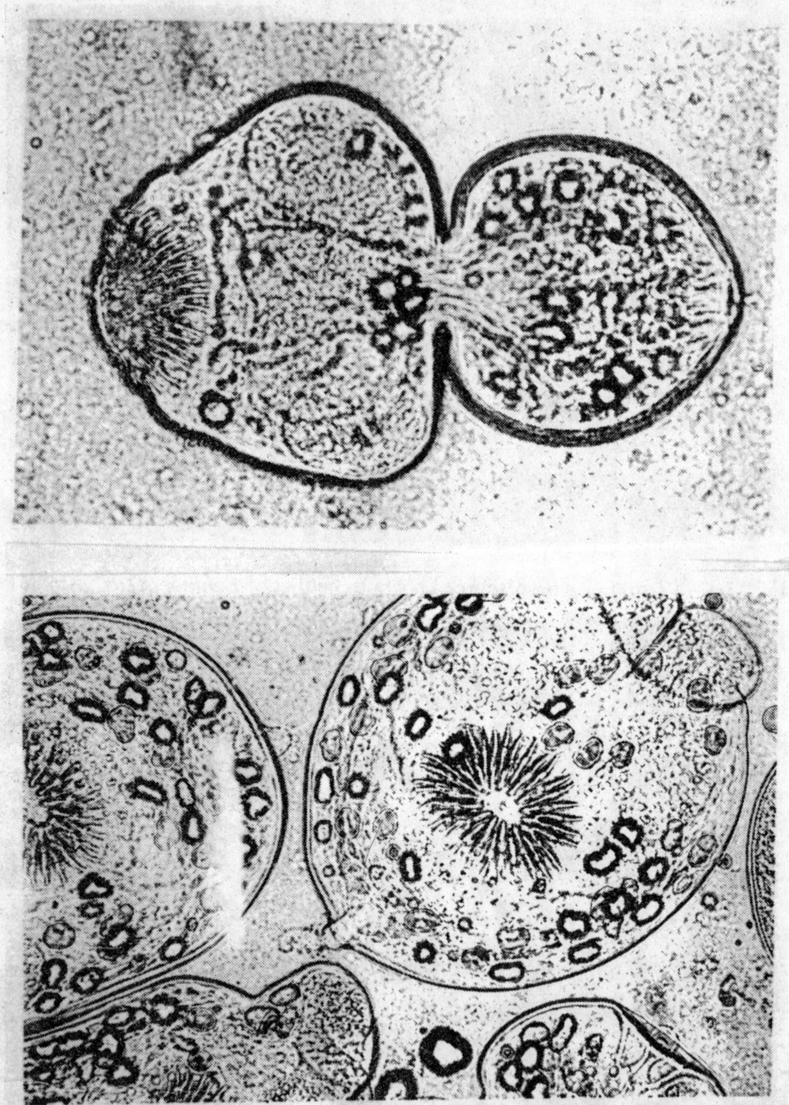

Fig. 12 Stages of evagination. (Note movement of hooklets from centre to exterior).

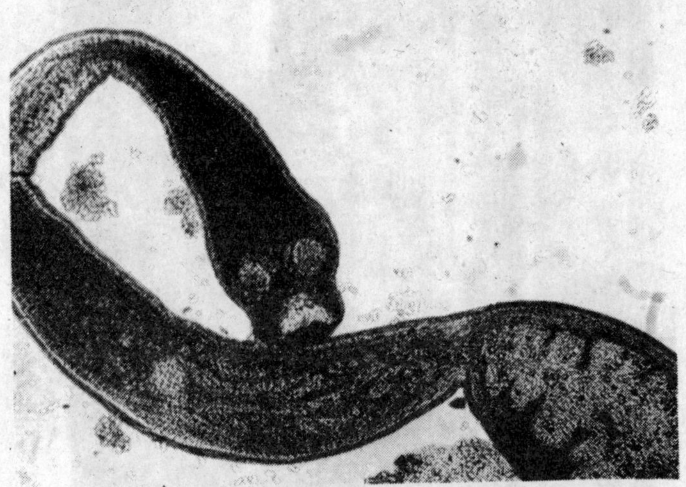

Fig. 13 Adult worm, *Echinococcus granulosus*.

Fig. 14 a) Heavily infected dog gut showing thousands of worms.

Fig. 14 b)   Moderately infected dog gut showing several worms (two A and B arrowed).

Life Cycles in Masailand

There are two distinct life cycles in Masailand:

a)   *The domestic cycle*

This is the one affecting man as suggested by Nelson and Rauch. Prior to that it was wrongly assumed that man got infected by eating and/or associating with wild animals. This domestic cycle is similar to the one in Turkana District.

b)   *Wildlife cycle*

This does not seem to be important in the spread of human disease (Chapter 1 Section III Sylvatic Cycle).

Section II

EPIDEMIOLOGY

There have been several epidemiological surveys in the past, mostly on animals, and other work involving observation of human cases seen at hospitals. Work by O'Leary, Nelson and French has shown that the northwestern part of Kenya, northwest Turkana, has the highest reported endemicity of the disease in man in the world. In that area 220 cases per 100,000 population per year are seen. While in the southern part, Masailand, the figure is less than 3 cases per 100,000 population per year. Kenya therefore has the highest endemicity reported in the world. The theories accounting for this are discussed elsewhere in this book.

Work by Okelo has shown that the disease also occurs sporadically in all other provinces in Kenya. Recent animal studies in Masailand by Eugster and later MacPherson and Karstad have shown that the disease is common in wild and domestic animals there, but the disease in man is of a much lower incidence than in Turkana district. The disease is thus present all over Kenya.

Formal surveys in man throughout the whole country using mass miniature radiography, ultrasonography and serology are needed. As part of the control measures in the Turkana district these measures are being applied particularly using portable ultrasonography and serology. Since the dog (Fig. 15) is the crucial link in the chain to human infection, its role has been considered in more detail elsewhere.

Definitive Host

Quote: "A man's best friend is his dog".

Author: Unknown

As has been discussed elsewhere in this book, the domestic dog is the most important definitive host of the parasite in man in Kenya.

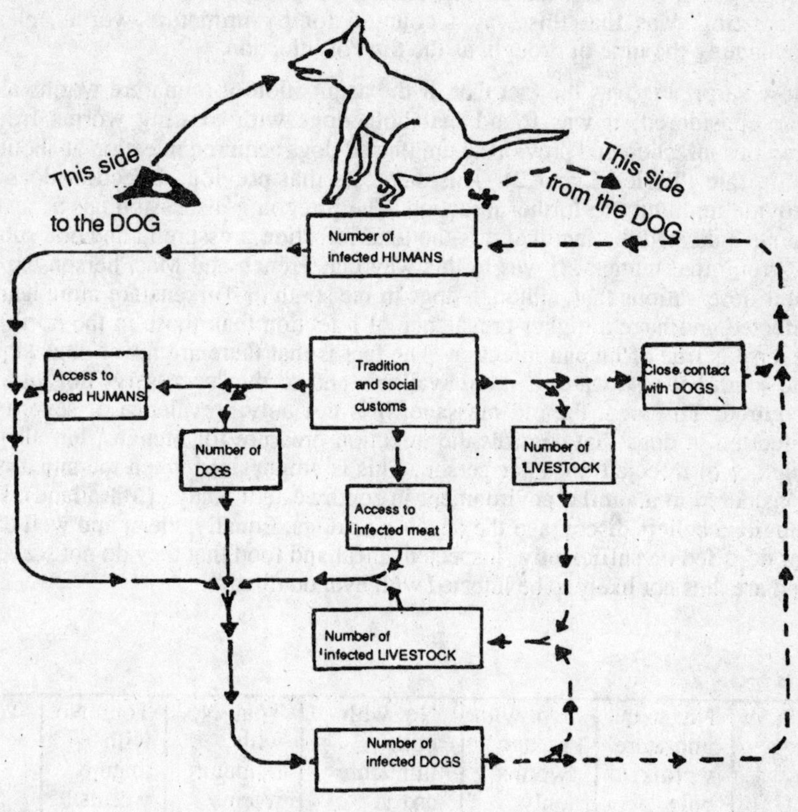

Fig. 15 Factors controlling the spread of hydatid disease with dog as control.

Effect of Drought

As a factor, the death of livestock during droughts had remained as a hypothesis until Ngunzi examined this in Masailand in 1984. His earlier observations had established a high prevalence of infection of dogs with *E. granulosus* before the drought. When many animals died and their emaciated carcasses (Fig. 16) were left in the open to rot, the prevalence almost doubled. What was even more interesting, was that this was accounted for by immature worms, clearly implicating the time of drought as the time of infection.

More surprising was the fact that, if the acquisition of immature worms alone was considered, it was found that both dogs with existing worms from a previous infection and previously uninfected dogs acquired infection at about the same rate (Tables 1 and 2). This suggests that previous infection does not provide immunity to further infection. The foregoing discussion has at several points indicated the fact that it is the total infection pressure in any one subject or group that matters. It was in this way that French and MacPherson explain their observations that, although dogs in the south of Turkana are more heavily infected and have a higher prevalence of infection than those in the north, the reverse is true of human infection. The fact is that there are fewer dogs kept in the south because water is more available and so the "necessary" dogs are not required. To repeat the old message, it is not only prevalence or severity of infection in dogs that controls the infection pressure for humans, but also the number of infected dogs per person. This is emphasized when the situation is considered in a similar environment in southwestern Kenya (Masailand). Here religious beliefs discourage the keeping of dogs. Equally, clean and well cared for dogs fed on sufficiently "inspected" meat and food that they do not scavenge and are thus not likely to be infected with hydatid disease.

No. of dogs	No. with immature worms only	No. with mature worms only	No. with both immature and mature worms	Total No. with immature worms	Total No. with mature worms	Total
92	13 (14%)	15 (16%)	6 (7%)	18 (20%)	21 (23%)	34 (37%)

Table 1   The prevalence of *Echinococcus granulosus* infections.

Fig. 16 Carcass of young cow that had died due to drought. Note scavenging of viscera.

	Total	Number with immature worms	Percentage with recent infections
Dogs with mature worms (previously infected)	21	6	29
Dogs with no mature worms	1	13	18

Table 2 Per cent recent infection with *Echinococcus granulosus* of previously infected and uninfected dogs in Kenyan Masailand between August and November, 1984.

Mathematical Models

The pioneering work of Anderson led to the idea that the behaviour of parasites could be expressed in terms of mathematical models. Three models representing the two extremes and the centre of a continuous spectrum of behaviour were produced.

1. *Under-dispersed*, represented by the positive binomial distribution. The parasite is evenly distributed among the hosts, each host having approximately equal numbers of parasites. This in its ideal situation, is a symbiotic relationship, such as exists between certain bacteria in the rumen of ruminants which digest the otherwise indigestible cell walls of the plants eaten. Here the host gains as much from the parasites as the parasite gains from the host.

2. *Random*, represented by the Poison distribution when the number of parasites per host is variable from host to host. Some hosts have no parasites, others have numerous parasites and the majority of hosts have an intermediate number of parasites. The number of parasites per host is a normal distribution, similar to height or weight distribution in a human population.

3. *Over-dispersed*, represented by the negative binomial distribution, clearly represents that of hydatidosis. In this the majority of parasites are confined to a few hosts that are heavily infected with many parasites.

Three different outcomes can be predicted. The first two mentioned here are lethal to the parasite, so its survival is impossible. First, the host is so successful that all parasites are destroyed. Second, the parasite is so successful that all hosts perish such that the parasite now having no hosts must also perish. The third, the usual situation with existing parasites, is that a balance is reached. The host population being reduced, but both parasite and host surviving.

The importance of these mathematical models is that they allow predictions to be made. In hydatidosis for example, once its prevalence is definitive, the secondary host population known and estimates of the infectivity at each transmission stage made, it is possible to compute the effect of any control measure. For example, it would be possible to predict the number of dogs that had to be killed to stop the cycle. Similarly the number of livestock not to be fed to the dogs could be calculated.

This technique has not as yet been applied to any disease, but with increasing availability of computers it will soon be. While this short introductory description has been based on these original calculations, the scope of models available is much larger, with more modern and recent ones offering even better predictions.

For the full mathematical analysis, readers are referred to the *Journal of Theoretical Biology*, 1980, Volume 82, pages 283-311. However, the following appendix and the graphs of Fig. 17 may help to illustrate the results of these calculations.

Appendix to Mathematical Models

Explanation of graphs in Fig.17.

Anderson, after a complex series of computations produces two basic equations:-

One for the changes in host population :

$$\frac{dH}{dt} = (r - \beta H) - \alpha P$$

The other for changes in the mean parasite burden :

$$\frac{dM}{dt} = M \left[\frac{\lambda H}{H-Ho} - (a + \mu + \alpha) \right]$$

Where

H   is the number of hosts

r   is the growth rate of hosts (i.e. the number born per unit time minus the number of deaths per unit time)

β   is the number of host deaths per unit time due solely to enviromental restrictions

α   is the number of host deaths per unit time per parasite. This represents the number of parasites dying because their host has died.

P   is the number of parasites

M   is the mean parasite burden i.e. $\frac{P}{H}$

λ   is the rate of production of transmission stage organisms by the parasite per unit time

Ho is the number of deaths of transmission stage organisms per unit time divided by the proportion of transmission stage organisms that eventually infect a host

a   is the number of hosts born per unit time

μ is the number of parasite deaths per unit time

$\frac{dH}{dt}$ and $\frac{dM}{dt}$ indicate changes of H and M with time.

Using these in the three distributions already mentioned, positive binomial, Poison and negative binomial, answers can be sought for various questions in these three different states of parasite dispersion.

For example the equilibrium or steady state may be sought by plotting graphs (Fig.17) for no change in time i.e.,

$\frac{dH}{dt}$ and $\frac{dM}{dt}$ are zero, that is:

$(r - \beta H)\ H - \alpha P = 0$ shown solid line and

$$M \left[\frac{\tau H}{H - Ho} - (a + \mu + \alpha) \right] = 0 \text{ shown dotted line}$$

In Fig. 17 this is done using different values for the variables to show the possible outcomes.

Underdispersed Positive Binomial

Graph *a*   Failure to co-exist. The lines do not meet.

Graph *b*   Stable co-existence at the point where the lines meet.

Graph *c*   Unstable co-existence possible at two points. The result is that these may oscillate and in the process cause either elimination of the host or parasite.

Random Poison

Graph *d*   Failure to co-exist similar to graph *a*.

Graph *e*   Stable co-existence similar to graph *b*.

             N.B. No unstable co-existence similar to graph *c* is possible.

Overdispersed Negative Binomial

Graph *f*   Failure to co-exist similar to graphs *a* and *d*.

Graph *g*   Stable co-existence similar to graphs *b* and *e*.

N.B Again no unstable co-existence similar to graph *c* is possible. Although these graphs are hypothetical, in a real situation the same is possible once the various statistics have been estimated. Thus host population reduction, mean parasite burden and environmental effect under stable co-existence can be calculated or derived from a graph. However, it is important to determine the relative proportion of these parameters in order to derive parasite failure.

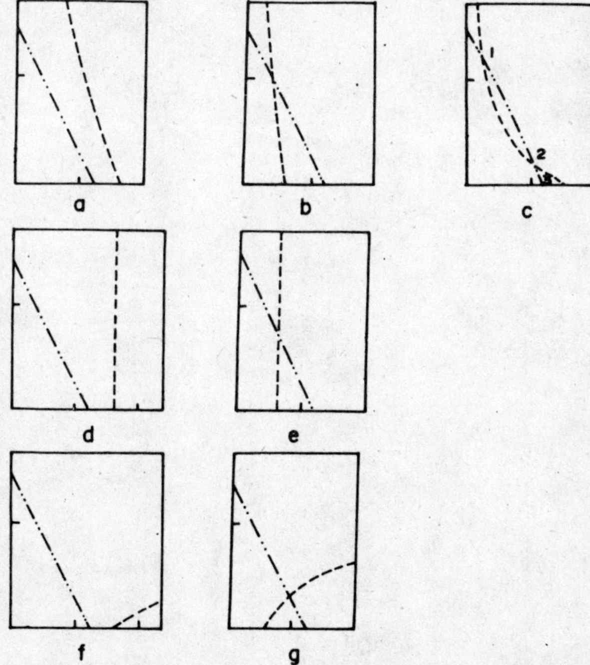

Fig. 17 Graphs demonstrating how a mathematical model can predict an equilibrium point. See appendix to mathematical models for explanation.

Chapter 3

Traditional Approaches

Section I

HYDATID DISEASE AND TRADITIONAL MEDICINE

It is not surprising that one of the oldest diseases to affect man should have aroused the interest of healers of all types in all generations. However, the greatest problem facing the "traditional healer" is diagnosis. "Western medicine" had this problem until the comparatively recent introduction of ultrasound in the 1950's. A swelling produced by a mass is obvious, but the ante-mortem determination of what the mass is, is not so easy. Hence the disease was usually only recognised at post-mortem when the classical fluid-filled cysts were seen. One of the earliest references to the disease is in the *Telmud*, the Jewish holy book. Interestingly, this also contains preventive public health measures such as the prohibition of association with dogs and eating pork (the source of a similar disease cysticercosis). From these origins arose two other holy books: the Old Testament Christian Bible where these prohibitions were omitted possibly because the diseases were unfamiliar to its European authors. The second, the Koran, where the prohibitions still persist probably because the disease associations were known to the author, Mohammed, who lived in Arabia. This is probably the reason why the incidence of hydatid disease in Lebanon is much higher in Christians than in Moslems.

The founder of "western medicine", Hippocrates, was probably the first to record diagnosis and treatment of the disease. In his aphorisms written on the Greek island of Kos around 400 B.C. he gives the name 'hydatid" from the Greek word for water " spos" (hydros), a very apt name for the fluid filled balloons that characterize the disease. He went further to describe a means of treatment using the hollow stems of goose quills (feathers) which were sharpened like a hypodermic needle and inserted into the cysts percutaneously. This procedure produced immediate symptomatic relief and often long term remission.

He did not record the expected side effect, anaphalaxis. This experience is similar to that of Cox in Kapenguria but differs from that of French in Lodwar where accidental needling subcutaneous cysts, mistaken for abscesses, did produce reactions. It is possible that the location of the cysts makes a difference and intra-abdominal cysts are less likely to produce systemic reactions than those in more superficial sites. This is supported in the next description of traditional surgical treatment of hydatid disease in Turkana.

Traditional Surgery

This is still practised in Kenya and shows a remarkable empirical knowledge of the effects of abdominal intervention. The patient is first made to lie on the side of the cyst for two days. After that an incision is made through the skin over the cyst while an assistant applies pressure to the abdomen so as to force the cyst to the surface. The incision is deepened until the cyst is reached and can be removed. The edges of the wound are then pulled together with reeds. The patient is not fed until a stool is passed. Although lacking in sterility and anaesthesia, the procedure accurately localises the cyst, avoids damage to other abdominal organs including the gut, minimizes blood loss and allows for paralytic ileus.

Traditional Scarification

This is performed for hydatid disease or any other internal mass (Fig. 18). In this procedure lines or rows of small incisions are made over the mass, particularly around its limits. The incisions are filled with charcoal, ochre and sand to cause scars. This treatment is often repeated, forming a series of scars marking the growth of the mass. If the time interval between these can be determined, which is usually about one year, then the duration and rate of growth of the mass can be determined. More than three sets often indicates hydatid disease since few patients with other tumours would survive for so long.

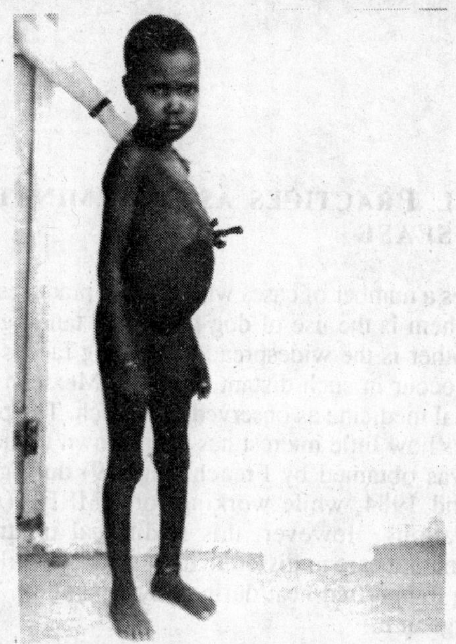

Fig. 18 Young patient showing traditional remedial scarification over a swelling caused by a hydatid cyst.

Traditional Herbal Remedies

The Turkana name for hydatid cyst is *lokamesipe*. A cyst is called *ekapespesit* and many concoctions are prescribed for masses in general. One of them is particularly popular in Turkana where hydatid disease is common. It is cut into small pieces, mixed with berries and pounded in water with stones. The resulting infusion is then given to the patients. There is no objective evidence for the efficacy of this treatment but its widespread use and popularity are indications that it deserves more study. Other forms of treatment by Turkanas include:

a) Taking herbs of *emus* and *emoja*.

b) The drinking of urine from sheep.

Section II

TRADITIONAL PRACTICES AS DISSEMINATORS OF HYDATID DISEASE

Schwabe describes a number of cases where social practices could spread hydatid disease. One of them is the use of dog faeces for tanning leather by Lebanese shoemakers. Another is the widespread use of dog faeces applied to burns and abrasions which occur in such distant places as Mexican Indian medicine and Turkana traditional medicine as observed by French. The paucity of reference to the latter indicates how little interest has been shown in the subject. Most of the data presented was obtained by French (Fig. 19) during his stay in Turkana between 1976 and 1984, while working for AMREF (African Medical and Research Foundation). However, this traditional treatment may not be as important as it first appears in dissemination since in Turkana the practice is to heat the dressing to near red heat during its preparation. The procedure would inactivate any eggs in it.

Fig. 19 *Lodoro* the potato-like root used as the basis for the traditional herbal remedy for swellings such as those caused by hydatid cysts in Turkana.

Chapter 4

Clinical Aspects and Diagnosis of Hydatid Disease

Section I

CLINICAL ASPECTS

The clinical diagnosis of hydatid disease depends on a high index of suspicion. This is especially important in areas with sporadic cases. The classical case of advanced abdominal hydatid disease usually looks well despite having extreme distension of the abdomen (Fig. 20). However, cysts appearing in rare sites such as in the orbit will often be missed. In the past enucleation of the eyeball was undertaken under the mistaken belief that it was due to neoplasm. Epilepsy may be the initial sign of intracranial hydatid cysts. A goitre with a cystic feel on palpation may be due to a hydatid cyst of the thyroid gland. The patient is usually euthyroid clinically. The vast majority of hydatid cysts are in the liver (about 70%) followed by the lung (20%). But Okelo has seen cysts in nearly every organ in the body.

Hepatic Cysts

Following the ingestion of the eggs by man, the liver is the first filter from the portal circulation. The masses feel cystic with gross but painless abdominal distension. The masses feel cystic and are typically non-tender. Some large cysts may contain several litres of fluid. Irregular hepatomegaly is usually present. Both kidneys may be enlarged and hydronephrosis may occur. The case shown in Fig. 22 was seen in 1986 and the hydronephrosis resolved following albendazole therapy. Some of the patients seen were found to have several litres of hydatid cyst fluid.

In early 1986 a young Masai female was admitted with acute abdominal pain. She had two very large cysts, one of which had ruptured. This is an unusual case. Another interesting case was of an adult Turkana female who had obstructive jaundice.

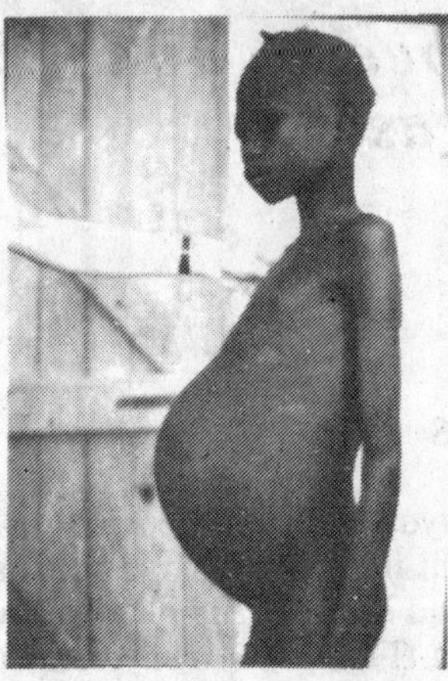

Fig. 20   Girl of about 12 years with large intra-hepatic hydatid cyst, later removed by surgery.

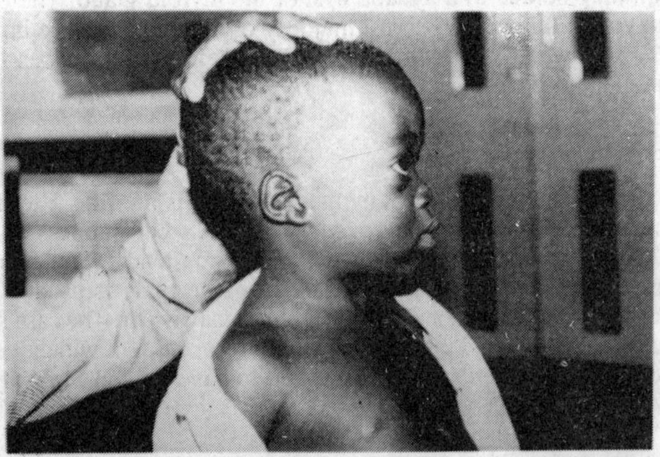

Fig. 21   A rare site for a hydatid cyst in the orbit.

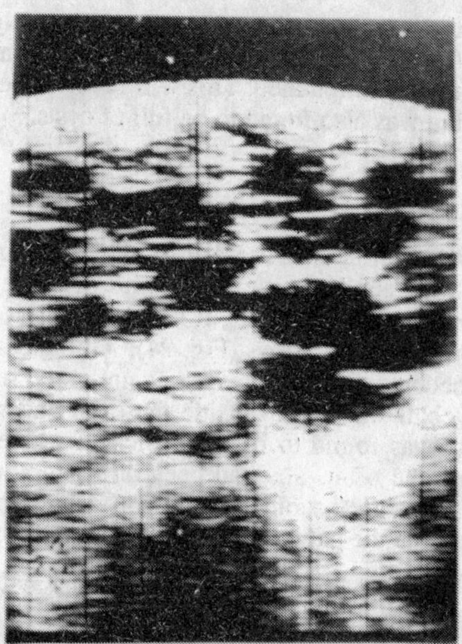

Fig. 22   Abdominal ultrasound picture showing a patient with hydatidosis.

Pulmonary Cysts

These account for about 20% of all our cases. They are usually asymptomatic, although the authors have seen pulmonary cysts as cough and haemoptysis. Pulmonary cysts may co-exist with pulmonary tuberculosis. An adult male nurse who had fever, cough, crepitations and the sputum expectorated, showed plenty of protoscoleces. The classical pulmonary cyst is easy to diagnose radiologically (Fig. 23). An interesting case was seen by Okelo in a child who had two large pulmonary cysts, one in each lung, both compressing the heart. The child preferred to rest in a head down position. Spontaneous rupture of a cyst during chemotherapy rarely occurs; but broncho-pulmonary distula may occur.

Bone Cysts

These are rare. A young female, referred to Okelo, had a spontaneous pathological fracture of the right tibia (Fig. 24). At operation, she was found to have multiple cysts at the fracture site. It healed on albendazole therapy. A young Turkana man was presented to French with a fracture below the left knee following a kick from a donkey. The fracture failed to unite and the unfortunate man had to have the leg amputated. This was before albendazole was available. In this case the tibia was also found to be full of cysts.

Orbital Cysts

A high index of suspicion is necessary to avoid unwarranted and unjustified enucleation of the eyeball. In the past some of the eyeballs removed were discovered to be due to hydatid cysts, at histology. Orbit cysts often appear with proptosis which may be very gross (Fig. 24). All the cases seen by Okelo have had gross papilloedema. The proptosis was unilateral in all of them. In one of the cases, the eyeball was right outside the orbit and when an operation was done the ectocyst was found to be very thick. It was excised and the eyeball replaced back into the orbit. She had been on chemotherapy with albendazole prior to surgery. The protoscoleces removed were all dead. Due to the long standing papilloedema the patient lost her sight from secondary optic atrophy.

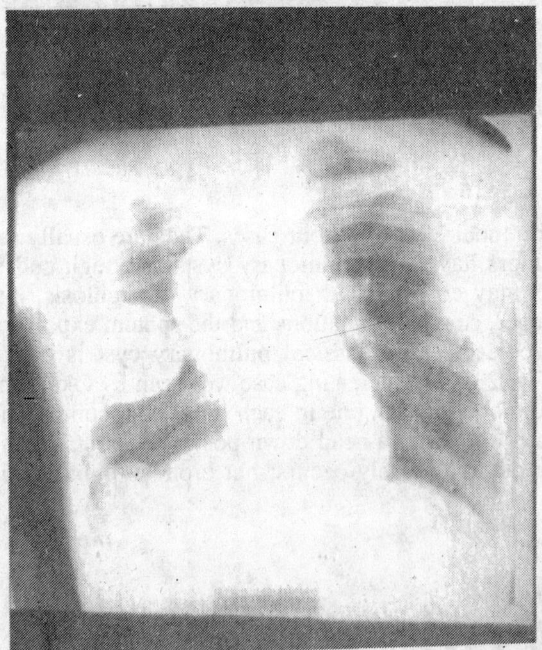

Fig. 23   Chest X-ray showing one large and several small cysts.

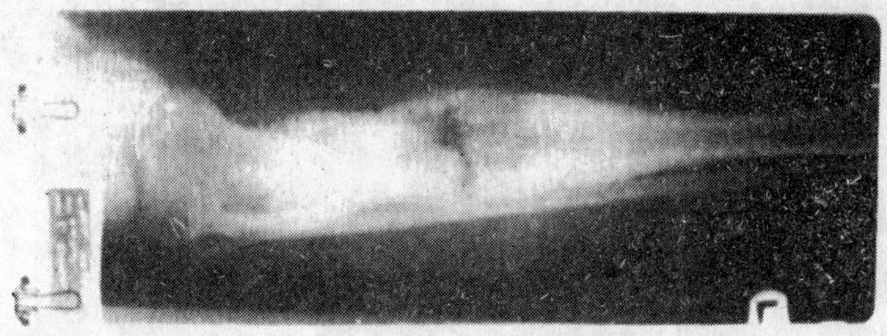

Fig. 24 Young female showing spontaneous pathological fracture of right tibia.

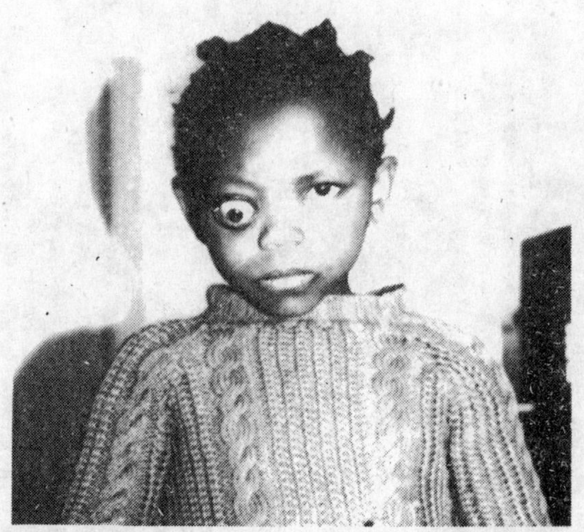

Fig. 25 Orbital cyst showing gross proptosis.

Cysts at Other Sites

In the last nine years we have seen cysts in nearly all organs of the body including skin, heart, pericardium, kidney, ribs, ovaries, urinary bladder, brain, post-operative scars, gall bladder, spinal cord, thyroid gland and old scars.

Section II

DIAGNOSIS

Immunodiagnosis

Diagnosis of most cestode infections involves parasitological examination of biological samples like stool for the presence of the parasite. Infection of man with hydatidosis represents the larval stage of the worm. The nature of this infection precludes parasitological diagnosis of the disease. Thus hydatid disease is diagnosed through procedures which rely on the detection of an immune response directed at the parasite, *E. granulosus*.

Parasite derived components with capacity to induce an immune response include the egg (oncosphere), protoscoleces and secretory/ excretory products of the germinal membrane. Hydatid cyst fluid is enriched with the parasite antigens and is commonly used as the source of hydatid antigen. A person infected with hydatid is sensitized or primed to the antigens which pass through the cyst wall and subsequently enter the blood circulation. This leads to the induction of anti-echinococcus antibodies and primed T-lymphocytes. A number of immunological techniques have been developed based on the detection of hydatid antibodies. The major specific antibody detected is IgG and IgA while IgM, are present in low rates. IgE levels are usually raised. Although testing for antibody has been the practice, major emphasis is now focused on methods designed to detect circulating parasite antigens or parasite specific immune complexes. This offers better advantages of distinguishing active infection from past experience or recovered cases.

In addition, monitoring the antigen is of tremedous value in evaluating a successful surgical operation and the efficacy of antihydatid chemotherapy. While the determination of T-cell function is an attractive diagnostic procedure, there are at present no reliable, reproducible or simple and rapid cell-mediated response tests for hydatid disease. Hence serological methods are the mainstay of hydatidosis diagnosis.

The most widely applied serological tests include:

- passive haemagglutination (PHA)

- immunoelectrophoresis, based on arc-5 band (IEP-5)

- enzyme-linked immunosorbent assay (ELISA)

- radioimmunoassay (RIA)

- double diffusion (DD)

- latex agglutination (LA) and

- complement fixation test (CFT).

Of these, based on the experience of the authors in Turkana District, the choice of method in the Kenyan situation comprises of IEP-5, IHA (PHA) and DD. With the use of highly purified *E. granulosus* antigens, ELISA systems should increasingly complement or replace some of these tests particularly in detecting and measuring circulating antigens, non-specific and specific circulating immune complexes.

Double Diffusion (DD)

This is one of the immunoprecipitation methods. Soluble antigen and antibody are allowed to combine in a solid support. Both diffuse towards each other and where the two meet, an immune complex is formed and precipitated if the proportions of antigen and antibody are optimal for cross-linking. The test is performed using purified agar such as Noble agar, iron agar and agarose in small petri dishes or on glass slides. Two wells are punched in the agar using DD template.

Hydatid cyst fluid antigen (HCF/Ag) is placed in one of the wells, while serum from a suspected *E. granulosus* infected patient is put in the other. Diffusion is allowed to occur in a moist chamber at room temperature. Between 24-48 hours later, precipitin arcs are formed if the serum contains anti-echinococcus antibodies.

The test is very simple to set up and reasonably cheap. Specificity to *E. granulosus* is enhanced by purification of hydatid antigens and both "arc-5" and "arc-4" precipitin bands can be demonstrated.

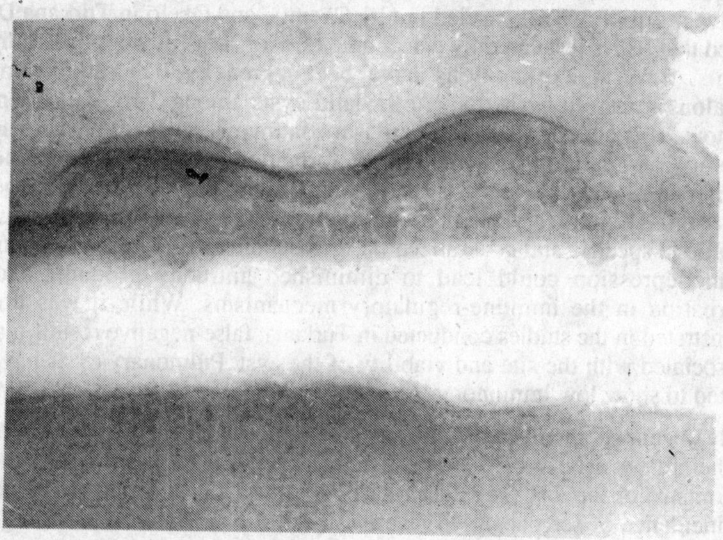

Fig. 26 Characteristic precipitation band designated "arc-5".

Immunoelectrophoresis (IEP-5)

Immunoelectrophoresis based on "arc-5" precipitin band is a modification of the conventional electrophoresis and immunodiffusion test. The test is performed with agarose on glass slides. Complex hydatid antigens are placed in wells made with the aid of IEP template and separated according to their overall surface charge, and molecular sized by applying a pontential voltage difference across the slide. Antigenic components thus migrate depending on their physical and chemical properties. After separation of the hydatid antigens, serum containing suspected anti-echinococcus antibodies is introduced into the trough made in the agarose. Immuno-diffusion is allowed to proceed in a moist chamber for 24-48 hours or more. Reactivity against the hydatid antigens is demonstrated by the formation of a characteristic precipitation band designated "arc-5" (Fig. 26). This is due to reaction between antigen 5 (one of the key diagnostic hydatid antigens) and specific *Echinococcus* antibodies. The precipitin band has a well defined morphology that is associated with hydatid infections and to a less extent cross-reactivity with cysticercosis and certain myeloma conditions. However, IEP-5 is considered the only reliable technique and provides a confirmatory test for *E.granulosus* infections.

A major drawback of the technique is a relatively high incidence of false negative reactivity. Work carried out by Chemtai and Okelo in Turkana District showed that IEP-5 test can only detect upto 60% of surgically confirmed hydatid patients. Several explanations have been given by the authors for this anomalous sensitivity reaction. As hydatid cysts among Turkana patients are enormous, it is conceivable that there is a continuous release of excess antigen into blood circulation. This leads to the formation of immune complexes and mopping up of anti-echinococcus antibodies leaving very low free antibodies in circulation. Alternatively, it is thought that infection with *E. granulosus* leads to profound specific and generalized immunosuppression. This parasite-induced immunodepression could lead to diminished antibody production due to perturbation in the immune-regulatory mechanisms. While it was not well demonstrated in the studies conducted in Turkana, false-negative results may also be associated with the site and viability of the cyst. Pulmonary cysts have been reported to show low immunoreactivity.

Finally, a general appraisal of serological tests reveals that conventional IEP has an inherent sensitivity of 5-10mg/dl which could also contribute to the poor performance of the test. Together, these factors account for the discrepancy of high incidence of sero-negative reactive hydatid patients detected using IEP-5 test.

Passive Haemagglutination Test (PHA)

This represents one of the most sensitive agglutination-based serological tests. It is carried out using human or sheep red blood cells. The former are preferred as they are devoid of heterophile antibodies which interfere with the performance of the test. Cells can be preserved with formalin or pyruric aldehyde to prolong their use.

Hydatid antigen is coupled chemically to the red blood cells. The coating is effected through treatment of the cells with either tannic acid, chromium chloride or glutaraldehyde. Excess antigen is washed off and the sensitized red blood cells are reacted with hydatid patients serum sample. Cross-linking of the antigen-coated red blood cells by the specific anti-echinococcus antibodies leads to agglutination or clumping together of the cells.

PHA is a highly sensitive test which detects more than 80% of surgically confirmed hydatid patients. This value is based on the experience of Chemtai and French with sera derived from Turkana district patients. However, specificity is dependent on the level of purity of the hydatid antigen. The use of the crude hydatid cyst fluid antigen gives false positive reactions against other taeniid worm infections, filariasis and fascioliasis.

The greatest application of PHA rests with its use in seroepidemiological studies. Sensitized red blood cells are lyophilised or freeze-dried and the test can be performed in the field without elaborate laboratory facilities.

Enzyme Linked Immunosorbent Assay (ELISA)

This is an enzyme immunoassay which has increasingly been developed and applied in the immunodiagnosis of parasitic diseases. The system is adapted for detection of specific antigen or antibody, specific and non-specific circulation immune complexes and rheumatoid like or anti-immunoglobulins. The test is performed in a variety of modifications. They all rely on the ability of either the antigen or antibody to adhere to plastic material e.g. polystyrene or polyvinyl chloride. After absorption of the protein, excess antigen is removed and the microtitre plates are incubated with enzyme-labelled antibody such as peroxidase-conjugated antihuman Ig (IgG, IgM or IgA). Excess conjugate is washed off and the plates are further incubated in the presence of an enzyme substrate. The product of the latter interaction develops as a colour that is visible or can be measured spectrophotometrically. The sensitivity and specificity of ELISA are high, but dependent on the purity of the antigen used.

The technique has been applied in the diagnosis of human hydatidosis and found to be particularly useful in the detection of circulating parasite specific antigens and immune complexes among free antibody false-negative reactive hydatid patients.

Altogether, it is widely recognized that no single test would be adequate in the diagnosis of hydatidosis. A combination of a battery of the immunodiagnostic techniques enhances both the specificity and sensitivity. It is recommended that at least three methods should be performed such as IEP-5, ELISA and PHA to satisfy all requirements for human hydatidosis confirmatory immunodiagnosis.

Ultrasonography

Abdominal ultrasonography has now been used successfully during the last nine years in the diagnosis and follow up of cases during chemotherapy. It is non-invasive and generally cheap after the initial installation costs. Ultrascans can distinguish solid from cystic masses. Typical hepatic cysts show up well on ultrascans (Figs. 26, 27).

Computerised Tomography Scan (CT)

Although computerised tomography scan is now available in Kenya, it is not yet widely used in most hospitals. When it becomes routinely available, the diagnosis and follow-up would be made much easier.

Chest Radiography

This is a very useful investigation in our experience. It should be done as a routine in all cases with abdominal cysts and pulmonary cysts. It is useful for follow-up during chemotherapy. They also detect the daughter cysts which give the characteristic "water-lily" sign (Fig. 28) on the chest radiograph.

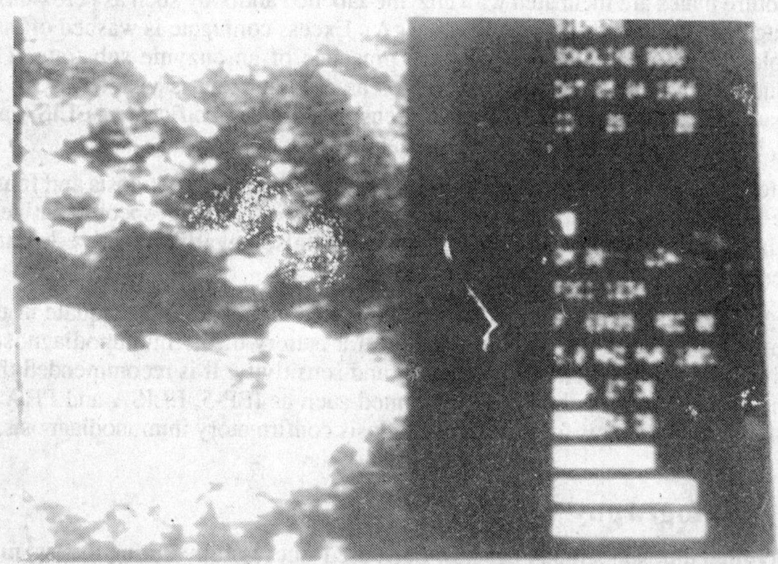

Fig. 27 Typical hepatic cyst shown on ultrascan.

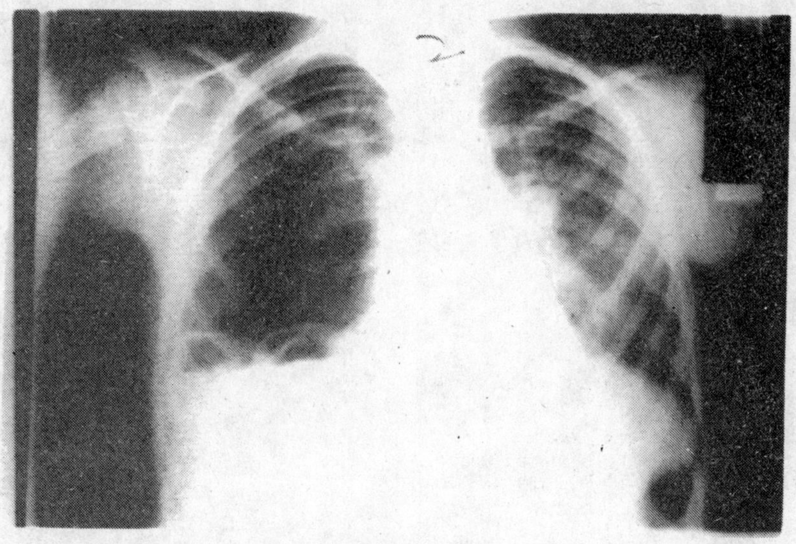

Fig. 28   Chest X-ray showing "water-lily" sign.

Chapter 5

Treatment and Control of Hydatid Disease

Section I

SURGICAL TREATMENT

For a long time this has been the only form of treatment available in Kenya, until 1978 when medical treatment became a possibility. Since 1976 about 100 cases have been operated upon each Year (Fig. 28). A review in 1983 in Turkana revealed the bad news that most of these post-operative cases were reappearing with recurrent severe and invasive cystic diseases. These recurrent cases are nearly always inoperable. Surgery should now be limited to well selected patients with large accessible cysts. Each operation should be preceded by a course of chemotherapy and after the operation another course of chemotherapy is necessary. No surgeon should today work in isolation from physicians when treating hydatid disease, since spillage and recurrence are almost invariable accompaniments of any surgical treatment. All surgeons should now use 0.05% cetrimide during swabbing of the abdomen, isolation of cysts and their subsequent excision.

Medical Treatment

The need for medical treatment in Kenya was realized in the mid 1970's. When *mebendazole* was first used in hydatid disease in North America and Europe in the late 1960's and early 1970's it showed a lot of promise. However, subsequent results in other parts of the world were not consistently satisfactory. In Kenya the first trial of its use was made in 1978 before a formal pharmacokinetics was undertaken. The initial trial using 40mg per kg of body weight in three divided doses daily for 8 weeks did not yield good results. A subsequent pharmacokinetics study by Okelo in 1984 showed, for the first time in Kenya, that *mebendazole* was effective, but its absorption was irregular and unpredictable. Only when a therapeutic level of 100ng per ml was achieved, and this requires fat to be given to the patient concomittantly, was the drug effective. The drug level was assayed by radioimmunoassay technique. Hence, *mebendazole* can be used effectively provided that fat is given simultaneously to improve its absorption.

Albendazole

The use of this *benzimidazole* derivative in hydatid disease was introduced in Kenya by Okelo in early 1983. To date, many cases have been successfully treated and are being followed up for recurrence. Currently the dosage used is 20 mg per kg of body weight in two divided doses given daily for eight weeks. Pulmonary cysts respond earlier than hepatic cysts. Fifty to 60 cases have been treated so far, and as yet no drug toxicity has been noted. Transient and early elevation of hepatic enzymes has been noted in two cases, but it tends to settle down quickly on its own. In general the drug is safe and well tolerated. Resolutions and cure may be seen on the chest X-ray (Fig. 31).

Combined Therapy

This is useful where surgery is going to be performed. A pre- and post-operative course of chemotherapy is advisable using *albendazole*.

Praziquantel

Recent reports in the literature indicates that this drug is effective against hydatid disease. Current work by Okelo in Kenya has shown that the drug kills hydatid cysts. So far several cases have been successfully treated.

Prognosis

Untreated hydatid disease is associated with a definite mortality which could be high depending on the location of the cysts. Pericardial, cardiac and pulmonary cysts are uniformly fatal if untreated. Cerebral cysts may die and calcify, and the only problem may be epilepsy. On the other hand continued growth of intracranial cysts, if unchecked by treatment, is fatal. The introduction of successful chemotherapy alone or combined with surgery has completely changed the outlook for these patients as has been shown in the previous sections. There is considerable morbidity associated with advanced hydatid disease. The greatly distended abdomen cannot allow the patient to do any work and his dependence on other people is inevitable. Untreated orbital cysts may result in blindness due to chronic papilloedema and optic atrophy.

Fig. 29 Operation showing removal of numerous daughter cysts.

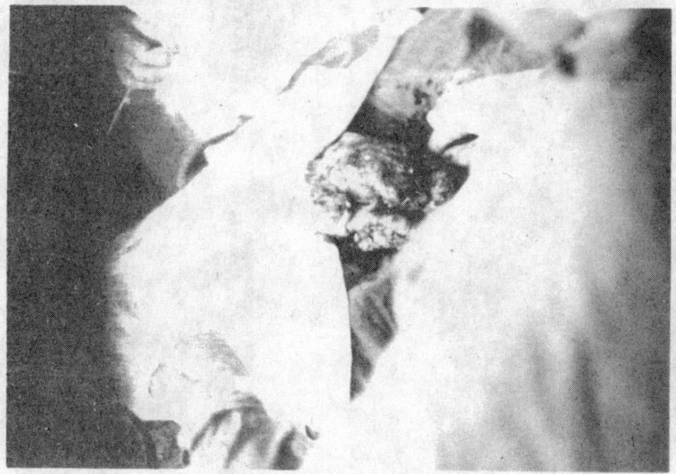

Fig. 30 Operation showing small intestine covered with thousands of small cysts, the result of spillage at a previous operation.

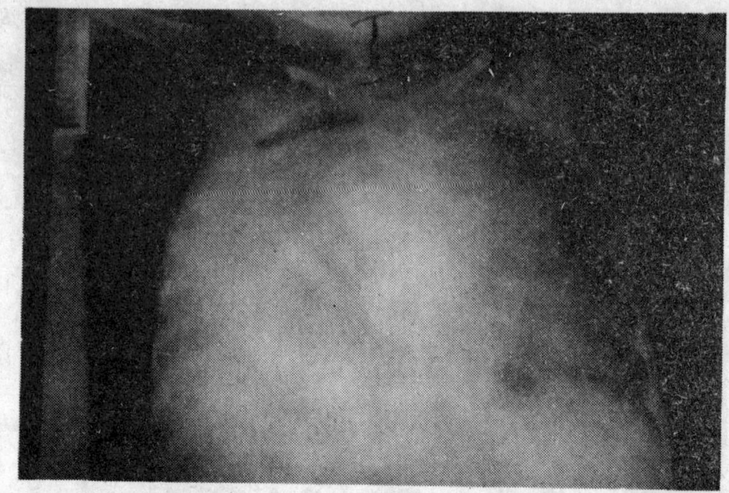

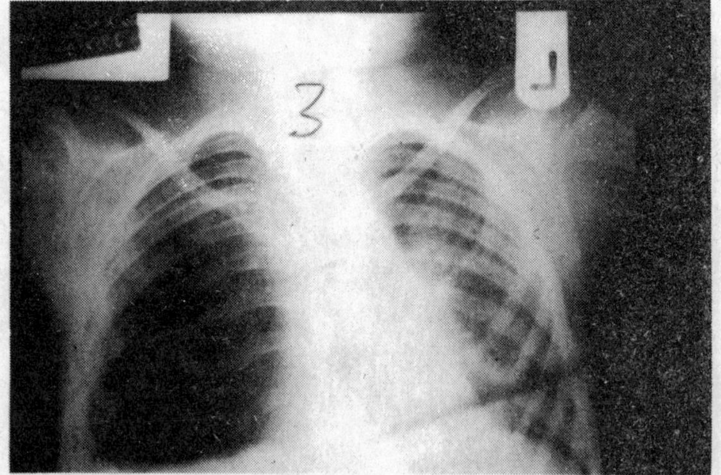

Fig. 31 Chest X-ray showing:
a)   large cyst in the right lung before treatment.
b)   complete resolution after treatment.

Section II

CONTROL

Human hydatid disease is an important public health and economic problem in this country. It is therefore understandable that the Government of Kenya found it necessary to start a Hydatid Committee in the Ministry of Health with members drawn from the University of Nairobi, the Ministry of Health, the African Medical and Research Foundation (AMREF), the Veterinary Department and the Department of Wildlife. At present the activities of the Committee are limited to Turkana District. The Committee has been in operation since 1983.

Its main objectives are:

1. Health education in the Turkana District as a long term plan.

2. Control measures regarding dogs in Turkana District.

3. Control measures against larval population in Turkana District. The plans that should eventually be worked out would be:

   a) Setting up of control authority.

   b) Personnel.

   c) A pilot study - now going on in the Lokichoggio area of the Turkana District.

   d) Health education programme in Turkana is going to be the ultimate answer and the final solution to the problem, but is inevitably going to be very slow because of a high rate of illiteracy and communication problems in a difficult environment.

However, it must be emphasized that hydatid disease is not just a problem of animals but also of man. Seven national control programmes in Argentina, Australia, Algeria, Cyprus, Falklands, Tasmania and Uruguay have certain components in common. These include reduction in dog population, regular dog dosing, upgrading abattoirs, dog registration and health education. These seven national control programmes have achieved some success. The only country in the world which has successfully eradicated the disease is Iceland. The major component of the Icelandic control programme was health education facilitated by the production of a small booklet written by a Dane in the local language. So every Icelandic family was able to read about the disease and to know the cause.

The dog population in Turkana has been reduced through an active spraying programme. Unclaimed dogs have also been destroyed. Previous dosing of dogs with *Arecoline* was found to be inefficient and erratic in action and a change to *Dronsit R.* was made. This contains *Praziquantel* which is a potent and useful antihelmintic drug in the dog as in other species.

The problem of setting up permanent structures such as abattoirs remains unresolved since the population is still largely nomadic. Permanent settlement is the aim of the control programme. These points illustrate the fact that although hydatidosis presents similar problems in different countries due to varying facilities, financial constraints and other conditions in each country, the priorities in implementing the control measures differ from country to country.

In concluding this section one point should be emphasized. The present control measures are still limited to only parts of one district. If it succeeds in that area it will then be extended to Masailand and the rest of the country since sporadic cases are increasingly seen in these areas. The medical treatment of cases now being successfully undertaken is part of the whole programme. This is because the incubation period is long and cases will continue to occur for many years long after transmission is stopped.

Chapter 6

Current Immunological Studies

IMMUNOSUPPRESSION

Various forms of specific or generalized immunosuppressions have been described in chronic cestode parasitic infections. This parasite-induced perturbation of normal immune responses occurs with hydatid infections. In the animal model systems, infection with either *E. multiloculalis* or *E. granulosus* leads to a depletion of thymus-dependent areas of lymphoid organs. Alteration in the functional composition of lymphocyte population and the production of immunosuppressive factors have also been reported.

Although there appears to be no clear documentation, it is conceivable that during *E. granulosus* infection, perturbation in immunoregulatory mechanisms could result in impaired T-cell-associated and antibody functions. With respect to T-cell functions, immunosuppression could reside in the alteration of T-cell subsets such as the elevation of T-suppressor mechanisms and depletion in helper T-cells as indicated by studies carried out by Chemtai. Impairment might also be linked to macrophage associated functions like antigen processing and / or presentation, phagocytic and oxidative metabolic activity. Evidence has been provided in other parasitic infections of alteration in lymphokine (cytokine) production. A similar defect could also apply in hydatid infections. One aspect of immunosuppression in hydatidosis studied by Chemtai is the suppressive activity of serum derived from *E. granulosus* infected patients. Experiments conducted have shown that sera from patients with hydatid disease are capable of suppressing normal lymphocyte proliferation (transformation) *in vitro* by up to 90% of normal values. The identity of the responsible suppressive factors has not yet been established. It is considered that high levels of circulating immune complexes, elaboration of cytotoxic antibodies and presumably auto-antibodies may exert the inhibitory activities observed. Furthermore, other factors such as prostaglandins, alphaglobulin fractions, alpha-feto protein and presently undefined parasite-derived components could suppress lymphocyte activity.

The authors are currently pursuing research directed at elucidating the immunological mechanisms involved in immuno-suppression in *E. granulosus* infected patients.

Hybridoma Antibody Studies

A normal individual when infected with *E. granulosus* responds to several antigenic determinants of the parasite. Several antibody species against each determinant are produced. While these polyclonal antibodies are specific, they are heterogenous and of limited value in the purification and characterization of hydatid antigens. It is important to have reagents which can be used to identify stage specific antigens and also distinguish the variants or sub-species of *E. granulosus*.

A novel approach was made by Kohler and Milstein. They developed an ingenious technique whereby a specific antibody of a normal cell can be rendered immortal through hybridization with a neoplastic cell. B-lymphocytes primed previously to an antigen are fused together with a selected myeloma cell line. The fusion is facilitated by a fusing agent, polyethylene glycol. Lymphocytes die in culture and the myeloma cell line also dies in a selected medium. Only the hybrids between the myeloma and lymphocyte derived cells survive in culture. The myeloma component provides the ability to grow in tissue culture and the B-lymphocyte contributes the functional activity of the immunoglobulin.

A large number of immunoglobulin-producing and specific antibody-forming clones are generated. Using hydatid antigens, selection of hybridoma-producing antibodies specific for *E. granulosus* can be established. Such antibodies designated hybridoma-derived monoclonal antibodies are of tremedous value due to the inherent advantages of high specificity, consistency and unlimited supply. Of particular importance is the generation of the hybridomas without the requirement of prior purification of the antigen.

Applications of Hybridoma-Derived Monoclonal Antibodies in Human Hydatidosis

a) *Identification of oncosphere:* All taeniid tapeworm eggs are morphologically indistinguishable. In epidemiological investigations and also in the control of hydatid disease, it is essential to clearly differentiate *E. granulosus* oncospheres from those of other worms. This has been achieved through an immunofluorescence staining technique using a monoclonal antibody which only recognizes the egg-derived oncosphere of *E. granulosus*.

b)   Purification and characterization of hydatid antigens: Monoclonal antibodies are also very sensitive tools for characterizing the complex hydatid antigens. Protoscoleces, germinal membrane and hydatid cyst fluid antigens are largely uncharacterized fully. Generation of specific monoclonal antibodies provide reagents that are used to define *E. granulosus* antigens of diagnostic and/or protective value.

At present there is a shift from antibody-based antigen dependent immunodiagnostic tests for human hydatidosis. Availability of well characterized anti-*Echinococcus* monoclonal antibodies should enhance both the sensitivity and specificity of serological tests designed to detect circulating hydatid antigens.

c)   T-cell activating antigens: Very often, it is difficult to discriminate antigens that exclusively or predominantly activate one compartment of the immune system. Highly purified hydatid antigens such as those obtained through affinity chromatography purification using *E. granulosus* specific monoclonal antibodies are particularly attractive. Co-culture of T-cells derived from *E. granulosus* primed individuals with such antigens, should provide information on the relative B-cell and T-cell reactivity to the antigens.

d)   Identification of variants or subspecies: Speciation is another complex issue in the epidemiological studies of hydatid disease. Apart from their use in purification and characterization of *E. granulosus* antigens, availability of the monoclonal antibodies would facilitate the genetic regulation of antigen production. Various species or subspecies of *Echinococcus* display different antigenic repertoires. *E. granulosus* specific monoclonal antibodies are useful in typing the variant sub-populations and hence solving the problem of speciation.

T-Cell Phenotypes

Different T-cell functions are associated with different T-cell sub-sets or sub-populations: helper/inducer T-cells (CD4), suppressor/cytotoxic T-cells (CD8), delayed type hypersensitivity T-cells (CDH$_{DTH}$) and macrophage activating T-cells.

In assessing the immune status of a person it is essential to examine the proportion of the T-cell sub-sets relative to the normal ranges in a given population of people.

Of particular significance in this respect is the ratio between CD4 and CD8 which is commonly used as an indicator of T-cell immunoregulatory defect. A depletion in CD4 or excessive increase in CD8 is often associated with immunosuppression or immunodeficiency problems during severe or chronic parasitic infections. Furthermore, the absolute value of CD4 is considered one of the key parameters in predicting prognosis of an infected patient. The relevance of these observations in *E. granulosus* infected patients is vigorously being pursued by the authors, in order to define more precisely the T-cell immuno-regulation. This could at least in part explain the mechanisms involved in parasite-induced immunosuppression.

Glomerulonephritis in Hydatid Disease

Immune complex mediated nephritis has been found in hydatid patients in Kenya. Studies are underway to characterise parasite specific antigens and immunoglobulins or complement components involved. The immunopathological processes responsible for kidney injury are also being investigated.

Pharmacokinetics of Albendazole

This is currently being looked into with a view to better understanding of therapeutic doses, drug metabolism, etc. in Kenyan patients.

Selected References for Further Reading

These references are selected from those available in the library of the Department of Medicine, University of Nairobi, and other references which those particularly interested should consult. For access, please contact the Chairman.

1. Allan, D., P. Jenluns, R.J. Conor and J.B. Dixon; 1981. A study of immunoregulation of BALD/C mice by *Echinococcus granulosus* during prolonged infection. *Parasite Immunology*. 311: 137, 1981.

2. Anderson, R.M.; 1980. Depression of host population abundance by direct life cycle macroparasites. *Journal of Theoretical Biology*, 82: 283-311.

3. Chemtai, A.K; 1980. Immunosuppressive effect of serum from human hydatid disease: Preliminary communication. *East African Medical Journal*; 57 : 11-15.

4. Chemtai, A.K., T.R. Bowry and Z. Ahmad. Evaluation of five immunodiagnostic techniques in echinoccossis patients. *Bulletin of World Health Organisation.*

5. Chemtai, A.K., P. De Baetselier and R. Hamers; 1984. Induction of protective immunity to *P. chabandi* by antigen-fed macrophager and antigen educated lymphocytes. *Parasite Immunology* Vol 6- 469-480.

6. Craig, P.S; 1986. Detection of specific circulating antigens, immune complexes and antibodies in human hydatidosis from Turkana (Kenya) and Great Britain, by enzyme-imraunoassay. *Parasite Immunology;* 8: 171.

7. Chemtai, A.K., G.B.A. Okelo and J. Kyobe; 1981. Application of immunoelectrophoresis (IEP) test in the diagnosis of hydatid disease in Kenya. *East African Medical Journal;* 58: 583-586.

8. *"Echinococcosis* and *Hydatidodis"* in *Tropical Geographical Medical* 2nd Edn. 1990 : McGrawHill.

9. Eugster, R.O; 1978. A contribution to the epidemiology of *Echinococcus hydatidosis* in Kenya (East Africa) with special reference to Kajiado District. *Thesis for the Degree of Doctor of Veterinary Medicine, University of Zurich.*

10. French, C.M., G.S. Nelson and A.M. Wood; 1982. Hydatid disease in the Turkana District of Kenya; I. The background to the problem with hypotheses to account for the remarkably high prevalence of the disease in man. *Annals of Tropical Medicine and Parasitology;* 76: 425-437.

11.   French, C.M. and G.S. Nelson; 1982. Hydatid disease in the Turkana District of Kenya. II. A study in medical geography. *Annals of Tropical Medicine and Parasitology;* 76:439-457.

12.   French, C.M. and W. Emonyi-Ingera; 1984. Hydatid disease in the Turkana District of Kenya.V. Problems of interpretation of data from a mass serological survey. *Annals of Tropical Medicine and Parasitology;* 78: 213-218.

13.   French, C.M; 1984. Treatment of human hydatid disease *Echinococcus granulosus* with mebendazole or surgery or both in the Turkana District, north west Kenya. *East African Medical Journal;* 61:113-119.

14.   Kungu, A; 1982. Glomerulonephritis following chemotherapy of hydatid disease with mebendazole. *East African Medical Journal;* 59:404-409.

15.   Macpherson, C.N.L., A.M. Wood and C.M. French; 1982. The use of cetrimide (R) as a scolicidal adjunct to hydatid surgery. *Proceedings of 3rd KEMRI/KETRI Annual Medical Scientific Conference;* pp. 21 - 35.

16.   Macpherson, C.N.L., L. Karstad, P. Tevenson and J.H. Arundel; 1983. Hydatid disease in the Turkana District of Kenya. III. The significance of wild animals in the transmission of *Echinococcus granulosus*, with particular reference to Turkana and Masailand in Kenya. *Annals of Tropical Medicine and Parasitology;* 77:61-73.

17.   Macpherson, C.N.L., A.M. Wood, C. Wood, C.M. French, L.O. Omondi, M. Mwangi, T.K. arap Siongok, P. O'Leory, M.M. Ngunnzi, G.B.A. Okelo and I. Mann; 1984. Perspective on options for the implementation of a pilot hydatidosis control programme in the Turkana District of Kenya. *East African Medical Journal;* 61:513-523.

18.   Macpherson, C.N.L., C.M. French, P. Stevenson, L. Karstad and J.H. Arundel; 1985. Hydatid disease in the Turkana District of Kenya. IV. The prevalence of *Echinococcus granulosus* infections in dogs and observations on the role of the dog in the lifestyle of the Turkana. *Annals of Tropical Medicine and Parasitology;* 79:51-61.

19.   Nelson, G.S. and R.L. Rauch; 1963. *Echinococcosis* infection in man and animals in Kenya. *Annals of Tropical Medicine and Parasitology;* 57, 136-149.

20.   Okelo, G.B.A; 1988. Mesangial proliferative glomerulonephritis in a patient with hepatic hydatid cysts : a case report in an African male. Transactions of the *Royal Society of Tropical Medicine and Hygiene;* 82 : 452.

21. Okelo, G.B.A. and J.A. Kyoje; 1981. Three-year review of human hydatid disease at Kenyatta National Hospital. *East African Medical Journal;* 58: 695.

22. Okelo, G.B.A. and A.K. Chemtai; 1981. Treatment of hepatic hydatid disease with mebendazole: Report of 16 cases. *East African Medical Journal;* 58: 608-610.

23. Okelo, G.B.A; 1984. Studies on human hydatidosis in Kenya. *Thesis for the degree of M.D. of the University of Nairobi.*

24. ———— 1986. Hydatid disease: Research and control in Turkana. III. Albendazole in the treatment of inoperable hydatid disease in Kenya - a report on 12 cases. *Transactions of the Royal Society of Tropical Medicine;* 80:193-195.

25. Okelo, G.B.A., A.K. Chemtai, S.K. Ongeri and D.K. Koech; 1987. Nephropathy in hydatid disease - a report on two cases. *Egyptian Medical Journal* (in press).

26. O'Leary, P., G.S. Gilchrist, C.M. French and M. Wood; 1979. Hydatid disease in Turkana. *Proceedings of the Association of Surgeons of East Africa;* 1:103-107.

[illegible]

[illegible]

[illegible]

[illegible]

[illegible]

[illegible]

Index